The Key to the Future

499

Highgate Literary & Scientific Institution

HIGHGATE Literary & Scientific INSTITUTION

Science Spectra Book Series

Series Editor: Vivian Moses, Queen Mary and Westfield College, University of London, UK

Volume 1
The Key to the Future
by John Cater

This book is part of a series. The publisher will accept continuation orders which may be cancelled at any time and which provide for automatic billing and shipping of each title in the series upon publication. Please write for details.

The Key to the Future

The Study of Earth History

John Cater PhD FGS

London and New York

First published 2002
by Taylor & Francis
11 New Fetter Lane, London EC4P 4EE

Simultaneously published in the USA and Canada
by Taylor & Francis Inc,
29 West 35th Street, New York, NY 10001

Taylor & Francis is an imprint of the Taylor & Francis Group

© 2002 Taylor and Francis

The right of John Cater to be identified as the Author of this Work has been asserted by him in accordance with the Copyright, Designs and Patents Act 1988.

Typeset in Optima by Wearset Ltd, Boldon, Tyne and Wear
Printed and bound in Malta by Gutenberg Press Ltd

All rights reserved. No part of this book may be reprinted or reproduced or utilised in any form or by any electronic, mechanical, or other means, now known or hereafter invented, including photocopying and recording, or in any information storage or retrieval system, without permission in writing from the publishers.

British Library Cataloguing in Publication Data
A catalogue record for this book is available from the British Library

Library of Congress Cataloging in Publication Data
A catalog record for this book has been requested

ISBN 0-415-27877-5 (HB)
ISBN 0-415-27876-7 (PB)

HIGHGATE LITERARY &
SCIENTIFIC INSTITUTION
No. 49972
Class 551 CAT
Supplier don Price – Date 03/03

For all the Steves
from John the Rock Doctor

CHUG!

Contents

		Page
Acknowledgments		ix
Chapter 1	What it's all about	1
Chapter 2	The study of Earth history	5
Chapter 3	Evidence of ancient environments	27
Chapter 4	Clocks of all sorts	61
Chapter 5	Events as time markers	73
Chapter 6	Snapshots from the past	101
Chapter 7	The big picture	123
Suggestions for further reading		129
Index		133

Acknowledgments

I want to thank Toine Wonders and Djin Nio for their contributions to this story. The opinions expressed are mine but much of the credit for its content is theirs. The faith and encouragement of my colleagues and friends has been invaluable – particularly from John Smewing, Alison Ries, Ed Bell, Bruce Sellwood and John Athersuch. The curiosity of friends like Steve Smith and Steve Cox inspired me to tell the story as I have. The book would never have been written without the love of Clare, Robbie and Lucy. The courage, honesty and generosity of my friends in far-off places has kept me working in trying field conditions. Everything would have been finished a lot quicker were it not for my commitments to Reading Motorcycle Action Group, 6-Pack, StrataData and ENRES International. So a big thank you to them all for making me take the time to mull it all over.

CONTACT: cater@ichron.com

1

What it's all about

Why study Earth history?

At first glance, the world seems a well-ordered and predictable place. The apparent order and stability of our day-to-day existence has allowed farming and civilisation to develop, regulated by the simple, customary changes of day and night, the tides and the seasons. At the scale of human lives, it seems that rivers flow steadily to the sea, mountains stay put at the horizon and sea level remains within the limits of the local tidal range.

It soon became clear to the earliest peoples that this vision of stability and predictability is dangerously unreliable. The most obvious departure from stability occurs because of changes in the weather. Extreme conditions are experienced only a few times in a person's lifetime: storms can drive the sea inland, flooding the coastal plains, while heavy rains cause rivers to break their banks and even change course, destroying flood-plain farms. These catastrophes are common enough that ancient civilisations learned to take precautions against them by building coastal defences, consolidating river banks or (more often) by simply avoiding flood-prone areas.

Bad weather is a minor problem in comparison with the truly rare disasters like the most powerful earthquakes and their associated tsunami ("tidal waves"), volcanic eruptions and major landslides. Such events are experienced more frequently in some parts of the world but generally affect us less than once per generation. Ancient peoples learnt to avoid living in the most disaster-prone areas but could only invent and pray to a pantheon of gods to try to avoid the unexpected. The ruins of Santorini and Pompeii witness the effectiveness of this approach.

On the longest time scale experienced by Man, the greatest disasters are recorded by distant tribal memories. Tales of great floods which destroyed whole civilisations probably refer to the rapid melting of continental ice-sheets which began about 12,000 years ago, since when world sea-level has risen by some 400 feet (130 m). Fossilised sea-shells are found on the highest mountainsides, suggesting still greater upheavals in prehistoric times. The rarest global catastrophes, the impact of large asteroids or comets, or the breakdown of the Earth's protective magnetic field, have probably never been experienced by humans.

To some extent the study of the Earth's surface, the processes that affect it and the history of such events has been driven by our need to understand and predict dangerous changes in our own environment. The second motivation has been to find useful natural resources, be they flint deposits or oil fields. But perhaps the greatest drive has been simple curiosity, the search for an explanation of our prehistory and for knowledge of the workings and history of our world.

Geology and religion: different views of natural history

The origin and history of the Earth and everything on it has fascinated people since the beginning of time and is inevitably bound up with religious ideas of creation. The very puzzling and often threatening phenomena which early people saw demanded explanations. But they had not yet developed the concepts of evidence and experiment and so had to use their imaginations to try to arrive at explanations.

Some of the ancient religions had views on the Earth's history which seem very strange to modern Man – the world was floating in a dish supported on the backs of a pair of elephants riding a vast turtle through the Void, or the Universe began as two twins in a wicker basket, or whatever. Although these myths can tell us a good deal about human mystical and psychological development, they contribute little to what we now call "Science". As our understanding of primitive human societies has grown, the meaning of historical legends has become clearer. There are, for example, many flood myths: the Norse ones tell of early times when the first gods slew the Frost Giant and survived the flood of his melting ice-flesh in a ship built from his bones; other cultures in China and Eurasia recalled similar flood-survival miracles; the Bible, of course, has the tale of Noah and his ark. These stories really may echo memories of ancient survival in the great floods affecting the whole globe at the end of the last Glacial Period; the story of the great Biblical Flood may originate in the catastrophic and sudden flooding of the Persian Gulf or the Black Sea basin.

Scientific ways of thinking and reasoning

Scientific and religious ways of thinking are different; whether science or religion is of more benefit to the individual and to society is a matter of one's objectives, beliefs and attitudes.

Religion demands commitment to and faith in a belief system, a set of defined axioms. It is therefore distinct from science because, unlike scientists, its adherents are usually unwilling to accept that their ideas may have to change as new facts are discovered. Some modern religions try to oppose "mainstream" science by arguing for their set of ideas in a scientific way. That fits in with scientific thinking as long as the proponents take into account real facts and are prepared to change their ideas if and when logic demands it. If not, they will be dealing with pseudo-science which takes them into untestable situations from which no further rational progress is possible.

Logical reasoning, the basis of all science and mathematics, is different: it aims to establish simple facts and build them up on a firm, verifiable basis, leading to more complex but sound sets of ideas. The procedure is simple. Ask a question – say, "what's that strange light in the sky?" Use your experience to conjecture some possible answers: it might be a flare hanging under a parachute, an aeroplane, a spacecraft driven by Superman or a Space Alien – or a weather balloon or satellite reflecting the sun. Choose the simplest possible solution which assumes the least number of unlikely circumstances (probably not the Superman option) and try to test it. If you can disprove it (the light is zigzagging wildly in the sky so it can't be a plane), move onto the next most likely answer. As Sherlock Holmes said, when you only have one possible solution left, then you have to go with that, no matter how unlikely it seems. This last remaining answer is called a *theory*. It survives until someone is able to disprove it. If they do, we have to come up with another alternative, not disproved by all the tests so far undertaken *but which might be disproved at any time in the future*. Scientists must be prepared to see their favourite theories destroyed in front of their eyes and cope with accepting someone else's ideas instead. It may be that we did not really understand one of the tests we did – returning to our example, maybe that strange light *was* a plane after all but one of its wings had fallen off and it was spiralling downwards out of control.

Our story is a scientific one about the Earth and its history, about what we think happened and why we think so. But remember that ideas change in the face of new evidence and some of the matters I shall describe will not look the same a century from now – perhaps not even next week. But this is where we stand now, and now is a particularly exciting time to be studying Earth history.

2

The study of Earth history

The most fundamental concept of Earth science is the realisation that Earth history is recorded by rocks, which are composed either of cemented and compacted sediment which in some cases has been chemically changed by heat and pressure during deep burial, or the crystallised products of previously very hot, liquid mixtures from the Earth's interior. The sediments record details of the conditions at their site of deposition, providing glimpses of an ancient world which can be as strange and exciting as any alien planet. More to the point, they record past changes in the environment which are the key to predicting our future.

Rock types

There are three basic kinds of rock – *sedimentary, metamorphic* and *igneous.* Sedimentary rocks are composed of compacted and/or cemented sediment. The composition of the original sediment is easily recognised in the resulting rock: *sandstones* are composed mainly of sand grains, *mudstones* largely of mud (i.e. silt and clay) and *limestones* mostly of calcium carbonate (originating primarily from the shells of ancient organisms). *Metamorphic* rocks form when sedimentary rocks are buried deeply enough to undergo chemical and physical changes, as in the formation of slate from mudstone. If the temperatures and pressures are high enough, the rock may melt to form liquid *magma,* which may be erupted at the surface to form *volcanic igneous rocks* (lava flows) or may cool and crystallise underground to give *igneous intrusions* (such as granite).

While the Earth sciences involve studying of all of these rock types, this book is concerned mainly with sedimentary rocks. This is because such rocks record most clearly details of past environmental change. They also contain economic oil and gas deposits, the search for which has led to many of our recent advances in understanding the Earth's history.

How old is ancient?

Earth history is measured in terms of thousands, millions and billions of years (a billion is a thousand million, 1,000,000,000 or 10^9). Exactly how we define the duration of events in Earth history is dealt with in later chapters of this book. For now it is important to understand that colossal lengths of time are involved. At the scale of every-day observation, it is very difficult to envisage how geological processes can act to deposit kilometre-scale thicknesses of sediments and build mountains. Simply stating that these things take millions of years does not convey the concept that gradual changes, like those we see on the banks of a river over a few years, are responsible for the large-scale evolution of the continents and oceans.

I remember reading an article when I was a boy which pointed out the scale of these numbers. Imagine a boy of 10 starting to count years at a rate of one a second for twelve hours of every day. He'd take just over half an hour to count back to the time of Christ. Ten hours later he'd be back to the end of the last Ice Age, when the great ice sheets of Eurasia and North America suddenly started to melt, flooding the world's oceans with meltwater and eventually raising sea level by 130 m (400 ft).

He'd take another two and a half of his counting days to reach back 125,000 years, through the last Glacial episode and a brief warm Interglacial to a similar period of melting of the previous Glacial ice sheet.

A little over six months later he would have reached 8 million years ago, when the forests of Africa began to give way to drier grasslands and his distant ancestors were encouraged to stand up and venture out onto the plains.

He would be fourteen before he had counted back to the extinction of the dinosaurs, 65 million years ago. He would have celebrated his (rather tedious) fortieth birthday well before he got back as far as 500 million years ago. At that time the most advanced life on Earth was simple shelled creatures such as clams, living only in shallow seas. There were no plants on land. The Earth itself is about 4,550 million years old (our young friend would be nearly 300 years old by the time he got there) and it formed midway through the life of the 12,000 million year-old Universe. That's definitely ancient!

With this understanding of the vastness of geological time, it becomes much easier to comprehend the long-term development of the Earth. However, it is not enough to understand that slow, everyday processes can do so much to change the shape of the Earth over vast periods of time. Unusual, catastrophic events have happened, and are waiting to happen again, which can do just as much overnight. These are not necessarily more exciting to the Earth scientist than slower, inexorable changes but they do capture the popular imagination. And no wonder – judging by past events, we could all be wiped out rather unexpectedly next week!

To understand the history of our planet, as recorded by the rocks, we need to know what processes affect the Earth's surface. We can then see how careful observations of modern processes, present rock outcrops and subsurface information can be used to work out the sequence of events which have shaped our world.

The law of Uniformitarianism

This basic concept states that "the present is the key to the past". This means that we believe that the same basic physical processes have been acting throughout Earth history – the force of gravity has remained much the same, water has had the same chemical properties throughout time, and so on. The result is that we should be able to explain our observations of what has happened by working out how such events could take place now. For example, if we observe isolated large pebbles of rock sitting in the muds of a deep ancient ocean, we ought to be able to explain how they got there by looking at modern situations where pebbles occur scattered on the muddy sea-floor. Such pebbles could be dropped by melting icebergs, they could be contained in the guts of dead and decaying sea-creatures (such as whales), or they could be caught up in clumps of vegetation drifting out to sea after a storm. We do not normally assume that they were formed by some process which cannot be observed today.

There is a limitation to this approach, which is considered again later in this book. Using the principle of Uniformitarianism in a blinkered way, we can overlook the effects of unusual events, of which we have little or no experience. Our pebbles could, for instance, be fragments of a meteorite which broke up above the ancient ocean and fell in separate pieces. This would be an unusual event, but still one which can be understood using modern understanding of the way nature works. We just have to be ready to imagine some pretty strange explanations for strange occurrences.

In some cases, our only record of unusual events comes from careful study of Earth history. Thankfully, we have no modern example of a giant comet impact such as the one which helped to wipe out the dinosaurs, or of dramatic climate changes which seem to have frozen the food in dying mammoth's mouths, or which allowed subtropical beetles to live in lakes on top of melting ice sheets. Studying such events can help us to understand the likely development of our planet from a wider perspective than that provided by looking at our short recorded history. This is what is meant by the phrase "the past is the key to the future" – an adaptation of the Uniformitarian principle which I have used in the title of this book.

The law of Superposition

Perhaps the most important concept in the study of sedimentary rocks is the realisation that new sediments are deposited on top of older ones. A pile of sediment builds up with the oldest deposits at the bottom. The layers of sediment are known as *beds,* or *strata. Bedding* (layering) is normally horizontal at the time of deposition, assuming that sediments were deposited on a flat surface. Where deposition has occurred on an uneven surface, for example on the edges of sand dunes, inclined or *cross-bedding* forms. This can be used to deduce the existence of ancient sand dunes (photograph 1). Flat layers of sediment beds may be tilted at a later date due to warping and buckling of the Earth's surface (photograph 2). The buckled layers may later be planed off by erosion, perhaps with younger flat layers deposited on top of them, forming an *unconformity* (photograph 3). The existence of such unconformities was seen by early geologists as evidence for ancient catastrophic upheavals of the Earth's surface. A major problem they faced was the time-scale involved; most early geologists were working within the received wisdom of the Church that the Earth is about 6,000 years old, so they had to fit all the observed events into that very short time span.

Photo 1 Cross bedding. Layers of sandstone near the pen were deposited on the sloping side of a sand dune in a river. The rippled surface above them was horizontal. Central Scotland.

Photo 2 Flat bedding surface tilted by compressional forces, western Pakistan.

Photo 3 Unconformity between tilted rocks and overlying near-horizontal rocks, which were deposited after the tilted rocks had been eroded. Central Turkey.

Sediment deposition and preservation

It is easy to understand that sediment deposition takes place at different rates in different places. Over most of the land surface there is little or no evidence of sediment deposition on a human time scale. We see some evidence of deposition on the flood plains of rivers and at coastal river deltas, lakes gradually silting up and beaches and spits building out along the shore, but relatively little changes in the course of our individual lifetimes.

The two requirements for sediment deposition are a supply of sediment and a sediment trap (such as a topographic depression) to be filled. Preservation of the deposit will only occur if there is no subsequent erosion.

Sediment is supplied to areas of deposition mainly by erosion of older rocks in upland areas either by wind, rain or ice action. Largely under the influence of gravity and atmospheric circulation, the eroded sediments are transported by the wind, rivers or glaciers. Additional sediment is supplied by the remains of animals and plants which are either transported by the same agents as the eroded sediments or form at the site of deposition. Since it is lower than the land, the most obvious final destination of these sediments is the sea. The majority of sedimentary rocks were indeed deposited in the sea. However, the surface topography of the Earth is constantly changing, with some areas subsiding while others undergo gradual uplift and erosion. Subsiding depressions form *sedimentary basins* in which sediment collects; some basins are flooded by the sea while others remain land-locked. *Subsidence* is the main mechanism for the creation of topographic depressions in which sediment can collect, the main driving force behind widespread subsidence of the Earth's surface being *plate tectonics.*

Plate tectonics

This fascinating and exciting view of the Earth in action is a fairly recent concept to geologists; few Earth scientists accepted it before the late 1960s. This is because plate tectonics provided a revolutionary new insight into the workings of the Earth (thereby forcing many scientists to abandon long-held views), and because the processes involved act at such slow rates, over such long time scales, that they are very difficult to demonstrate at our scale of observation.

We now know that the interior of the Earth is made up of extremely hot rock which wells up at local hot spots like a pot of extremely viscous, boiling fluid. This slow-motion "boiling" process, termed *convection*, affects the surface in two ways. Firstly, in ancient times (before about 3 billion years ago), lighter material accumulated as a sort of scum floating on the surface of the molten Earth and gradually collected together to form the continents. Most of the continental crust which is now at the surface had accumulated

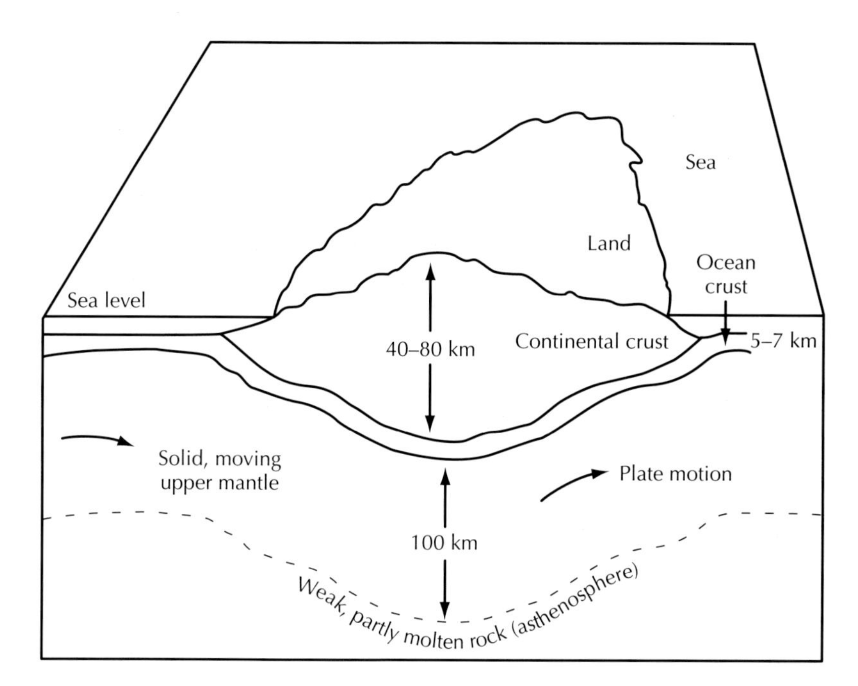

Diagram 1 A "raft" of continental crust floating on the solid upper mantle which moves on a weak, partly molten layer called the asthenosphere. The crust and upper mantle form the lithosphere, which is divided into rigid plates over 100 km thick.

by about 2 billion years ago. The continents are rafts of less-dense material which float on the moving subsurface and have been carried around on it, in a slow-motion shuffle called *continental drift* (diagram 1).

The surface of the denser moving material grows a solid (but flexible) oceanic crust, forming the ocean floor in which the continents are embedded, sticking out like cherries from a fruit cake. The oceanic crust is exposed between the edges of the continents, where it forms the floor of the oceans. The boiling convection motion of the underlying interior causes the rigid sections of ocean-floor crust, normally about 5 km thick and known as *plates*, to move separately from one another. The continental crust, averaging about 40 km in thickness, rides on top of the oceanic plates as if on a raft. Because the continental crust is embedded in underlying ocean-floor material, the continents drift around the Earth's surface, carried along by the motion of the underlying plates. The movement of the continental rafts is slow – about 5 cm per year, which is equivalent to 50 km per million years.

The motion of the plates is caused by the upwelling movement of hot material from deep in the Earth's interior. Hot rock rises to push the surface plates apart above the area of upwelling. The hot rock cools and solidifies to form kilometre-scale wrinkles, or ridges, of newly-formed ocean-floor crust, mainly along the mid-lines of the world's oceans. These "mid-ocean" ridges are where the plates well up and separate. The new ocean-floor crust, of course, takes up the space made available as the plates separate. This separation is possible because at the opposite end of the plate, where it meets a neighbouring plate, its edge sinks below its neighbour into the Earth's interior, in a process called *subduction*. As it sinks into the hot interior, the cool oceanic crust is partially melted, generating earthquakes and volcanoes. Patterns of earthquake zones and volcanoes delineate the edges of the plates, and these show where new crust is being formed by constant eruption at mid-ocean ridges while old crust is being consumed and melted at identifiable *subduction zones* (map 1 and diagram 2). Basically, the Earth can be seen as a vast rocky engine, continually converting heat from the interior into the motion of the surface plates.

Eventually, the continental rafts reach subduction zones where two or more oceanic crustal plates are colliding. Being less dense than oceanic plate material, the continents cannot be drawn down into the Earth's interior. Instead, the oceanic crust of one of the plates sinks below the edge of the

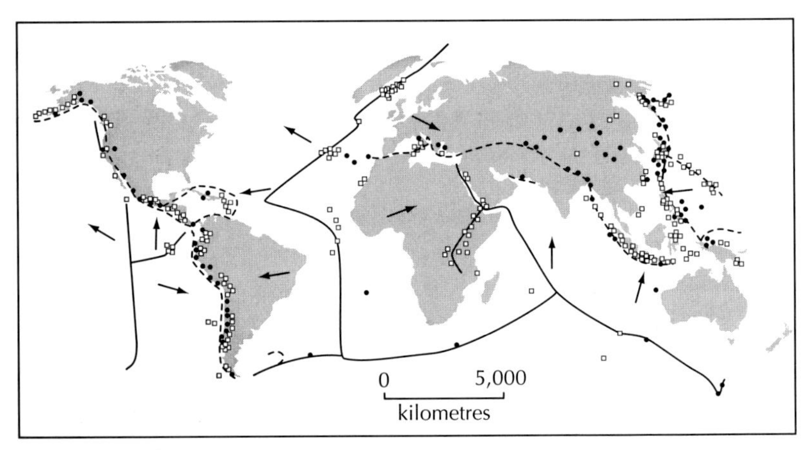

- ▫ Active volcanoes
- • Earthquakes over 7.8 on Richter scale since 1900
- —— Plates growing at "mid-ocean" ridges
- ---- Plates being destroyed at subduction zones
- ◄— Plate movement

Map 1 Earthquakes and volcanoes are concentrated at the edges of plates, either where they are created at "mid-ocean" ridges, or where they are destroyed at subduction zones.

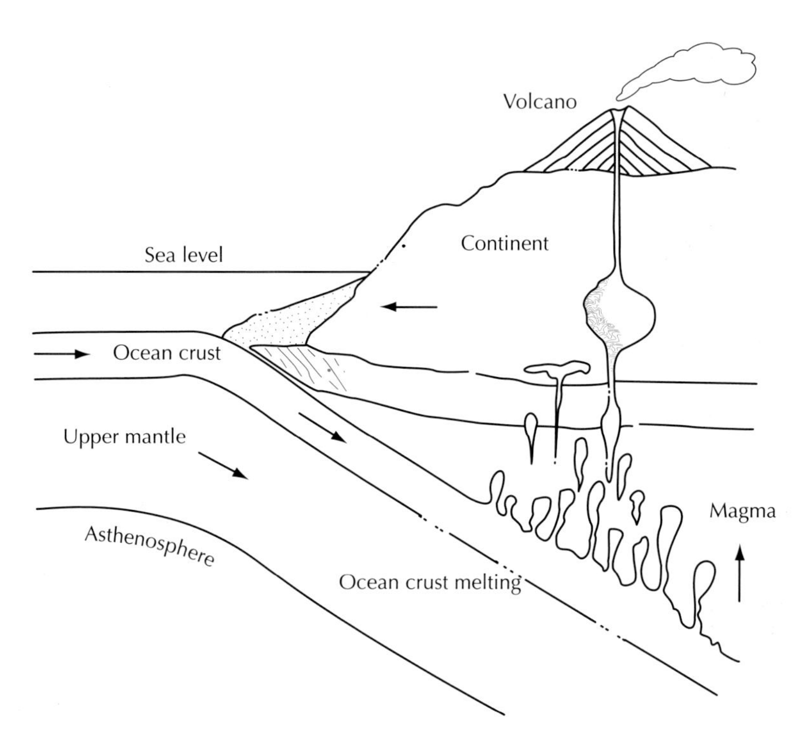

Diagram 2 The lithosphere is destroyed at a subduction zone, where oceanic crust is melted at depth to form liquid magma which rises and erupts from volcanoes. Continental crust is too buoyant to be subducted and remains at the surface.

continent which is being carried on the other, and active volcanoes erupt through the continent as magma from the partially-melted oceanic crust rises from below. The volcanoes of the Andes and the western Pacific are mainly of this type. In time, a second raft of continental crust may arrive at the subduction zone, again riding on the back of the oceanic crust which has been diving under the first continental raft. This causes continental collision. Neither continent can be drawn down into the earth's interior because, thanks to their low density, they are too buoyant. At first, the pressure of collision deforms the leading edges of the two continents, forming mountains as the crust buckles and folds. The rate of convergence is slow – on the scale of a few cm per year – and the resultant uplift of mountain belts occurs at similar rates. Eventually, the ocean crust breaks behind one of the continents and is drawn down into the interior, forming a new subduction zone while the welded continental fragments remain in place, joined at a mountain range which records their collision. New volcanoes erupt at the edge of the continent, where the new subduction zone is generating freshly melted magma.

The boiling motion of the Earth's interior varies through time so that the rate of upwelling changes gradually, altering the rate of movement of the plates: during the earlier history of the Earth, over a billion years ago, the rates of motion were probably much faster than at present. But the rate of motion is unlikely to have changed quickly over the last billion years except at local areas of unusual activity. Moreover, the oceanic plates sometimes split apart as new patterns of upwelling develop beneath them to generate new spreading centres and new mid-ocean ridges. Rarely, new spreading centres may form underneath areas of continental crust; as a result, the continental crust splits apart, and a new ocean forms between the two diverging sections of continent.

The constant creation and destruction of the oceanic plates means that oceanic crust is recycled over long periods of time; the oldest known region is about 200 million years old. Continental crust cannot be recycled by subduction although the edges of continental plates melt and recrystallise due to heating during collision processes. The oldest known continental crust is found in the stable interiors of continents and is over 3,500 million years old, with some recently dated fragments being perhaps 4,000 million years old.

The gradual but fairly constant movement of the continental crust causes the slow deformation of the surface, giving rise to sedimentary basins and uplifting highland areas. Relative motion of the continental crust is accommodated by plastic warping and folding, or by brittle breakage. Breaks in the crust are *faults* and sudden episodes of movement along such faults cause earthquakes. Minor earthquakes also result from the movement and eruption of magma, allowing limited prediction of volcanic eruptions by monitoring earthquake activity below volcanoes. Other earthquakes result from vertical movements of sections of the crust, for example due to the melting of thick ice caps and the subsequent "rebound" of formerly depressed land masses as the weight of the ice is relieved. Major earthquakes occur at the edges of sedimentary basins where parts of the crust are subsiding due to plate tectonic forces.

Types of sedimentary basin

The types of sedimentary basin can be understood by considering their relationship to the plate tectonic forces which created them. Their detailed classification is a complex subject but it is based on simple fundamental principles which are easy to understand in the light of plate tectonic theory.

There are three types of plate tectonic motion which affect the continental crust – *collisional, divergent* and *lateral*. That is, fragments of continental crust can collide either with each other or with separate oceanic crustal plates, they can be pulled apart and drift separately on different oceanic plates, or they can slide laterally past other plates. In most cases, collision or separation is oblique and involves some degree of lateral motion.

When continents collide, the colliding edges tend to fracture along gently-sloping *faults*, called *thrusts*. Relatively thin slices of crust (a few kilometres thick) ride up along the thrusts over the adjacent land surface. The slices bend and buckle internally, but tend to remain intact. Long-term collision builds up a wedge-shaped pile of these slices, with new slices either riding up "piggy-back" over the earlier ones or breaking forwards into previously undeformed rocks at the front of the wedge. Collisional mountain ranges such as the Rockies and the Himalayas accumulate as a stack of thrust slices. The thrust stack behaves a bit like a pile of rubble pushed in front of a bulldozer; as it builds up, the stack can collapse under its own weight, forming small, unstable internal basins (photograph 4).

More importantly, the weight of the thrust stack depresses the undeformed edge of the continent, causing subsidence all along the edge of the rising mountain belt. This forms a *foreland basin*. Such basins occur in front of all major collisional thrust stacks, including the Ganges Basin south of the

Photo 4 The rocks on the right of this valley in Zanskar (northern India) have dropped down along a major fault zone, along the line of the river, by several kilometres relative to the High Himalayan rocks to the left (south), due to gravitational collapse during the uplift of the Himalayas.

Himalayas and the Po River Basin south of the Alps. As continental collision progresses, new thrusts may cut through the sediments deposited at the edge of the foreland basin, incorporating them into the growing thrust stack. These sediments are generally tilted up and rapidly incised by rivers flowing from the mountains, providing excellent exposures of rocks deposited in foreland basins throughout the world. Many of the outcrop examples shown later in this book are from uplifted foreland basin deposits.

These processes of continent–continent collision are quite different from those operating when continental crust collides with oceanic crust. We have already noted that the oceanic crust normally loses the battle against gravity during subduction and is sucked down at the collision zone. The edge of the continental crust is heated, partly by friction against the subducting oceanic plate, but mainly by the upwelling of magma from below as the subducting oceanic plate is partially melted during its descent. Volcanoes erupt through the continental crust but there is little subsidence close to the collision zone. Instead, the crust is heated, and new crustal rock is formed by the cooling and crystallisation of igneous intrusions. The crust close to the collision zone therefore tends to swell up, forming a ridge of continental crust alongside the collision zone itself. Sediment eroded from this ridge collects in the subduction zone and is deformed by friction against the subducting oceanic plate.

Japan is a good example of such a setting (diagram 3). The main islands consist of old continental crust and to the east is the Pacific Ocean plate which is moving westwards towards mainland Asia. The oceanic crust is being sucked down under Japan, diving below the continental crust just to the east of Japan. Volcanoes such as Mount Fuji erupt through the crust. Sediment eroded from the uplifted land mass collects in the deep trench which marks the collision zone. The line of Japanese islands is arcuate in shape; similar arcs of volcanoes can be seen all along the western edge of the Pacific Ocean, marking similar collision zones as the Pacific Plate collides with adjacent oceanic and continental crust.

The deep basins which lie between these volcanic arcs and the oceanic plate are called *fore-arc basins* because of their position in front of the volcanic arcs. The sediments which collect in them are deposited in deep water – sometimes many kilometres deep. Those sediments are unlikely to be lifted up to the surface unless continental crust arrives at the collision zone, riding on the back of the oceanic plate. This will not occur in the foreseeable future in the Pacific (probably for tens of millions of years) because no continental crust is being carried westwards on the Pacific Plate towards Asia. However, it has happened elsewhere in the past. One of the best-known examples is in southern Scotland, which once was separated from

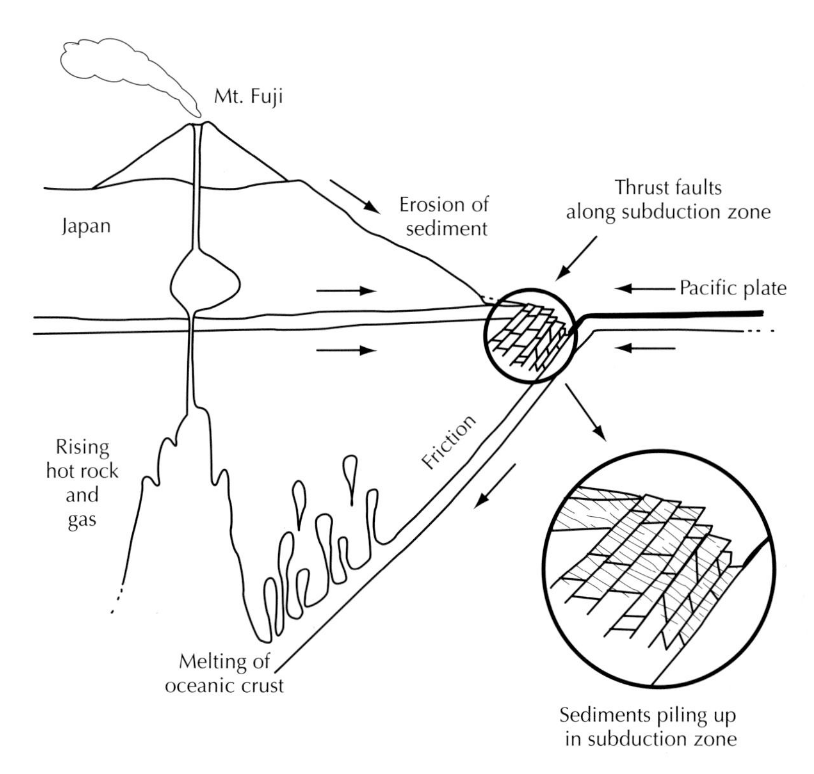

Diagram 3 Sketch of the subduction zone east of Japan. Melting of the Pacific ocean crust generates the lavas of Mount Fuji. Sediment eroded from the land is piled up as it collects in the oceanic trench at the subduction zone.

England and mainland Europe by a wide ocean. Some 450 million years ago, this ocean was being subducted northwards below Scotland. Deep-water sediments were collecting in a fore-arc basin, in what is now the Southern Uplands. They were uplifted to the surface during the collision of Scotland with southern England about 50 million years later. Fore-arc deposits are also well exposed in the Makran region of western Pakistan where the Arabian Sea is being subducted northwards below Asia.

Volcanic arcs do not stay put over long periods of time. Instead, they tend to migrate slowly, either towards or away from the interior of the adjacent continent. This is due to changes both in the rate of subduction of the oceanic plate and in the steepness of the dipping oceanic plate as it dives under the adjacent continent. If the volcanic arc moves towards the continental interior, the edge of the continent is compressed and thickened, forming a range of mountains and volcanoes along its length. This process was responsible for the formation of the Andes, for example. In much the same way as in foreland basins associated with continent–continent collision zones, basins form on the interior edge of coastal mountain belts due to the loading of thrust sheets; they are then known as *retro-arc basins*. If, on the other hand, the arc migrates away from the continental interior, the edge of the continent is extended and subsides. This forms a *back-arc basin*, illustrated by the Sea of Japan located to the west of the islands, a broad area undergoing gradual subsidence. The subsiding areas are separated by steep faults which break up the basin into a pattern of individual sub-basins, each with a different subsidence history and different pattern of sediment fill. The Aegean Sea east of Greece, another example of this type of basin, is being created during the subduction of the southern Mediterranean under Europe, as Africa moves northwards. The volcanic arc runs through Cyprus and the southern Cyclades islands, including the resort island of Santorini. Volcanic activity there is now rather unimpressive but a colossal eruption in about 1,900 BC destroyed a civilisation flourishing locally and is thought to have been the origin of the legend of Atlantis.

Although dense oceanic crust usually is subducted below continental crust at collision zones, in some cases a slice of the oceanic crust may become separated from the main thickness of the down-going oceanic plate during collision and be pushed up onto the adjacent continent, by a process called *obduction*. This probably happens most often where an old mid-ocean ridge reaches a subduction zone, with part of the upstanding ridge being sliced off at the subduction zone and obducted. Slices of ancient oceanic crust are fairly common at old collision zones. Good examples occur in Cyprus, Greece and Turkey, although probably the best (and prettiest) is along the north-eastern coast of Oman. There, about 90 million years ago, the oceanic slice pushed a wedge of deep-sea sediments in front of it, up onto the continental crust of Oman. These sediments record deposition in sea-water thousands of metres deep during the Mesozoic Era, between about 200 million and 100 million years ago. They are now exposed in deep valleys eroded through the kilometre-scale slice of oceanic material, which is called an *ophiolite* (an old geological term derived from the Greek for "snake", since the weathered oceanic rock is patterned like snake-skin in places). The ophiolite weathers in the harsh climate into amazing spires and pillars, resembling the cover of a science-fantasy book. However, the science facts derived from study of such regions tell us the history of sediment deposition on the continental margins in the aeons before collision occurred, in particular giving insights into processes acting in the deep sea which are almost impossible to observe at present because of the water depths involved.

Divergent basins form when fragments of continental crust are pulled apart. The most important type is the *rift basin*, formed in the early stages of continental separation and the creation of a new ocean. The upwelling of hot material in the Earth's interior creates new oceanic crust, as we saw earlier. The pattern of upwelling changes through time as new upwelling cells are formed deep in the Earth and old ones become inactive. If a new upwelling zone forms below an existing continental plate, the continent is first uplifted above the rising oceanic crust and then splits apart to form rift valleys. Volcanoes erupt along the valleys, fed by magma rising from deep in the Earth's interior. Lakes and rivers first occupy the valleys until the continent splits apart and the valleys are drowned by the sea. Eventually, as the continental fragments drift apart, new oceanic crustal material erupts along the valleys and a new ocean is born. Strangely enough, it seems that our own species was also born in such a rift valley, the Great Rift Valley of eastern Africa.

The Great Rift Valley marks a split in the African continental crust filled by lakes and rivers. Volcanoes including Mount Kilimanjaro occur all along the line of the rift. This rift is about 20 million years old but has not yet developed to the stage of oceanic crust formation and possibly never will. Instead, beginning about 25 million years ago, a wider rift formed further north-east along the Red Sea and Gulf of Aden. About 15 million years ago the margins of south-western Arabia were uplifted by 2–3 km and huge quantities of volcanic lava were erupted in Yemen and north-eastern Africa. Rift valleys formed along the lines of the Red Sea and Gulf of Aden, with lake and river sediments collecting in them in much the same way as in the Great Rift Valley today. But Arabia became separated from the African continent about 10 million years ago and oceanic crust was created – in a thin strip in the southern Red Sea and a wider band along the Gulf of Aden. This spreading zone also failed to develop fully because Arabia had already collided with Asia. As a result of this collision some 15 million years ago, mountains formed in Iran and Turkey, setting up a contest between the spreading forces trying to separate Africa and Arabia and the resistance at the collisional zone in the Zagros and Tauride mountains.

Due to cooling and contraction of the crust on the edge of the new ocean, complete separation leads to subsidence of the formerly uplifted edge of the continent. The continental margins bordering the Atlantic Ocean illustrate this late stage of rift development. The Atlantic formed by continental rifting between about 150 million and 50 million years ago, beginning in the south with the separation of Africa from South America and ending with the separation of northern Europe from Greenland. It has been widening ever since, with oceanic crust being formed along the entire length of the mid-Atlantic Ridge. The continental margins have since cooled and subsided,

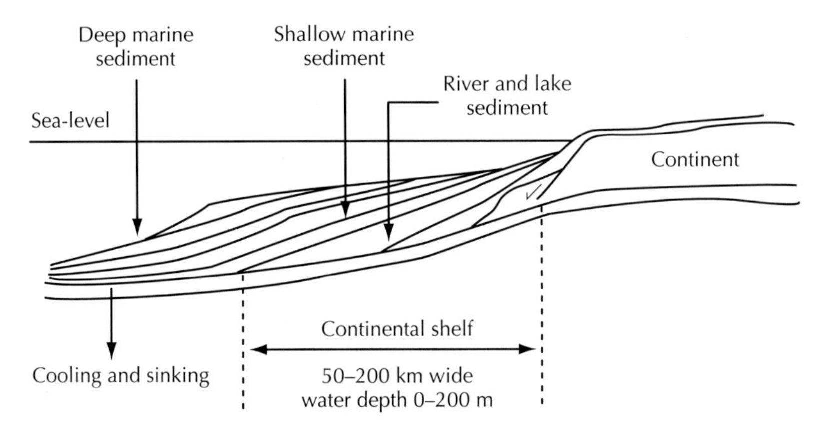

Diagram 4 Passive continental margins subside as oceanic crust cools and sinks with age. The continent develops a fringe of shallow marine sediment building out over older, subsided river and lake sediments, known as the continental shelf. The width of the shelf mainly depends on how much sediment is eroded from the continent.

forming *passive margin basins* which have filled with marine sediments overlying the sediments deposited in the lakes and rivers of the original rift basins (diagram 4). *Passive margin sediments* consist of continental shelf deposits as well as muds and sands deposited in the deep ocean; the latter are commonly seen preserved as rocks only where continents have collided and, as a result, are generally exposed as highly deformed fragments of the original sediment systems. More complete, less heavily deformed examples are sometimes exposed in areas that have undergone obduction (see above).

When continental crust is uplifted and split in the early stages of rift formation, the split tends to form along three main rifts. Two of them eventually widen to form new oceanic crust while the third is generally abandoned, forming a *failed rift basin*. Failed rifts are quite common along the margins of the continents, extending inland away from the growing ocean. The basins usually contain sediments deposited in lakes and river systems, possibly with some shallow marine sediment on top, deposited as the sea flooded the subsiding failed rift valley. Major rivers draining the continental interior tend to flow along the valleys to the sea and form large deltas, as illustrated by the Niger delta today. The North Sea is a failed rift system and many of its oil fields are located in sediments deposited by similar deltas.

Failed rifts have an early history of rapid subsidence along steep faults, as the continental crust is pulled apart during continental separation. A pattern of rapidly-subsiding sub-basins develops, similar to that in back-arc basins but with potentially greater subsidence rates. Uplifted blocks shed sediment into deep lakes or seas during this period of rapid subsidence. Once the rift is abandoned, subsidence is much slower, resulting from widespread sagging of the crust as it cools. The post-rift phase of subsidence creates a wider *thermal sag basin* overlying the older, narrower, failed rift basin. Thermal sag also affects the interiors of large continents after an episode of heating or compression, as in some of the large interior basins of Asia and North America.

Lateral plate motions, where one plate is sliding past another, can create narrow basins due to local pulling apart of the crust. This occurs if the fault line separating the two crustal fragments is not straight, with pull-apart occurring at *releasing bends* in the fault-line. Relatively small but rapidly-subsiding basins of this type occur along the San Andreas Fault zone in the western United States. *Compressional deformation* occurs at *restraining bends* along the same margin. Changes in the relative direction of plate movement can also cause more widespread extension or compression. North-west Pakistan has a long history of extension and compression along its margin with Iran and Afghanistan due to the continents sliding past each other until India collided with Asia to form the Himalayas. This history is recorded by sediments deposited in extensional basins along the western edge of Pakistan which were later uplifted during compression along the same margin.

The time-scale of basin development

The major driving force behind sedimentary basin development is plate tectonics. We know from present-day measurements at spreading margins (such as the middle of Iceland) that oceanic crust moves across the globe at rates of a few cm per year. Studies of the ocean crust have shown how old it is at numerous locations throughout the world, revealing a pattern of spreading and migration away from the mid-ocean ridges and eventual destruction at the subduction zones. The age of oceanic crust increases away from the spreading ridges, showing how rapidly the crust has spread in the past (map 2). This shows that past spreading rates were about the same as today, generally ranging from 2–15 cm per year. This is equivalent to 20–150 km per million years, with a rate of about 50 km/million years being about normal for plate tectonic movement.

This general rate of change gives an idea of the overall rate of development of sedimentary basins. The rates of subsidence of basins depends on the mechanisms by which they form. Extension at releasing bends along translational fault zones causes the most rapid subsidence. Subsidence rates of several km per million years are recorded in small basins along the San Andreas Fault; in effect, the bottom drops out of such basins because the

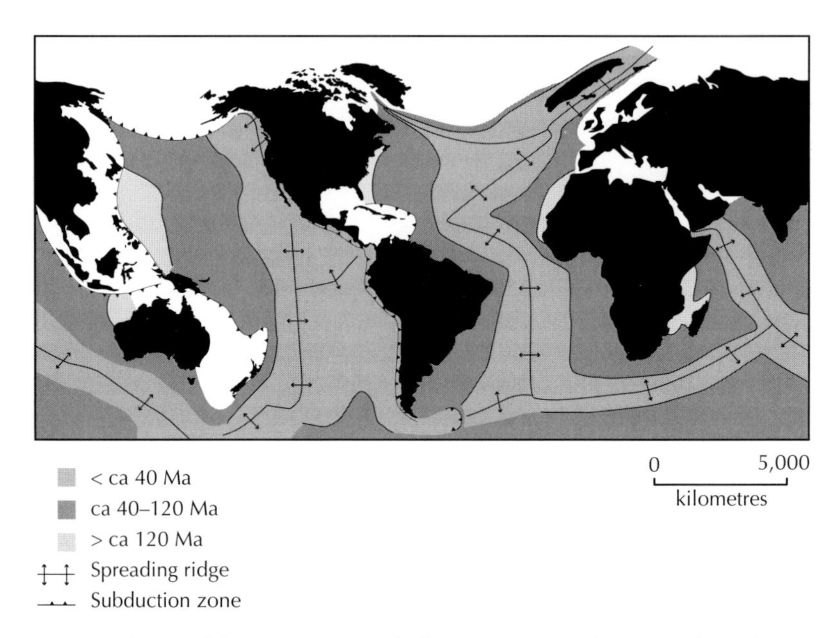

Map 2 The age of the oceanic crust, which is younger away from spreading ridges.

underlying continent has split apart along a narrow line of weakness. Rifted continental basins subside less rapidly because extension is distributed across a wider area as a series of separate fault blocks. Rapid subsidence may occur for short periods above one fault block before the main location of subsidence shifts to another fault block. In the Red Sea and Gulf of Suez, for example, only 2–3 km thickness of sediments accumulated in the rift basin as a whole during its 15 million-year evolution before the onset of sea-floor spreading, but each local area collected most of this sediment thickness in much shorter periods of time.

The Red Sea appears to be spreading rather slowly, each side moving away from the mid-line at a rate of about 1 cm/year. Thus, the 100 km-wide strip of oceanic crust present in the southern Red Sea was formed mainly during the last 5 million years, although the oldest oceanic crust in that area is about 10 million years old. Spreading during the first 5 million years must have occurred at very slow rates. Based on the 25 million year age of the oldest rift-valley sediments found in drill cores, the rift valley stage of development of the Red Sea basin, before oceanic spreading, lasted some 15–20 million years.

Thermal sag of cooling continental crust, whether in continental interiors, above failed rifts or at passive continental margins, is significantly slower than rift-related subsidence along faults. It also slows down with time as hot crust cools more rapidly and sags more quickly than cooler crust. The cooling crust at separating continental margins subsides at about 100 m/million years, slowing down rapidly after the first few tens of millions of years. Thermal sag in continental interior basins is slower, about 10 m/million years.

Continental collision, also driven by plate motion, is about as fast. India has migrated northwards on the back of the Indian Ocean Plate at 5–10 cm/year (50–100 km/million years) and is now crushing northwards against Tibet. Collision and mountain building have occurred in this area throughout the last 45 million years, during which time India has moved 2,000 km northwards at an average speed of 5 cm/year. The main phase of Himalayan uplift began about 25 million years ago and rapid uplift of the Tibetan Plateau began about 15 million years later. Tibet now has an average height of 5 km, although migration of land mammals across the plateau was still possible until about 2 million years ago. The main range of the Himalayas continues to rise at 2 mm/year (2 km/million years) but this is counteracted by erosion so the resultant elevation of the mountain peaks is changing much more slowly, if at all. Major subsidence of the Ganges foreland basin began about 15 million years ago and, at about 500 m/million years, more than 6 km of sediments have accumulated in the basin since then. As the Himalayan thrust belt migrates southwards (or as India migrates northwards, depending on your viewpoint), the rate of subsidence has increased with time, bringing any point in the basin gradually closer to the point of maximum loading by the thrust sheets. At about 20 m/million years, the average subsidence rate of typical foreland basins is much slower, partly because of averaging out very slow initial subsidence rates and eventual very rapid rates, and partly because of the smaller thrust loads in most collisional mountain belts compared to the exceptionally large-scale situation in the Himalayas.

Overall, then, basin subsidence derives from plate tectonic effects which last for tens of millions of years and result in the continental crust dropping several kilometres at 10–1,000 m/million years. Subsidence rates vary significantly between different types of basin and within individual basins.

Sediment deposition through time

We have already noted that the two requirements for sediment deposition are a supply of sediment and a topographic depression to be filled, while preservation of the deposit will occur only if there is no subsequent erosion. The formation of various types of subsiding basins provides the topographic depressions in which sediments can collect: *transported* sediments and *indigenous* sediments. The former are mainly carried by the wind or flowing water. They usually consist of some mixture of mud (grains less than 0.06 mm in diameter), sand (grains 0.06–2 mm in diameter) and gravel (grains over 2 mm in diameter). The relative mud/sand/gravel composition of sediment in a basin depends on the type of rock being eroded in the highlands, the processes affecting the sediment during transport to its point of deposition, and the processes going on after deposition. Both the distance from the erosional source (grains get worn down during transport) and the flow power of the transporting agent (weak flows cannot transport large grains) are important. *Indigenous* sediments include coal (which forms in swamps from the burial and compaction of plants), salt and other evaporated crystalline deposits, and the shells and organic remains of animals and plants which lived in the basin. Thick deposits of indigenous sediments accumulate only in places where transported sediments are not being deposited. Of course, indigenous sediments may well be eroded and transported after deposition, forming another type of transported sediment.

As basins develop, the kinds of sediment deposited in them tend to change due to changes in the local environments as well as modifications in slope between the basin and adjacent highlands. Following the basin classification described above, it is easy to envisage the changes in sediment fill that may occur during the development of a continental rift (diagram 5). During the initial rift valley formation, volcanic lava flows and muds deposited in lakes form the dominant fill, together with local gravelly deposits formed by landslides and fast-flowing streams at the steep edges of the rift. As the sea gradually invaded the rift valley, salt deposits may have formed during periods when the sea was temporarily cut off from the ocean and evaporated. Coal swamps may also have formed at that time. As the two continental fragments separated, sandy river deposits formed along the new ocean margins, passing offshore into more muddy marine sediments which accumulated further from the supply of sediment. The shallow sea was home to abundant marine organisms, mostly microscopic in size, and their shells built up to form limestone deposits along the continental margin. Finally, the passive margin subsided into deeper oceanic water, where very fine muds were deposited, together with rare subsea avalanches of sand transported from shallower water areas.

This is the kind of general picture of a basin's evolution that can be developed by studying the sequence of sedimentary rocks which were deposited in it. To do this from basic observations of rocks at outcrop requires a thorough understanding of the processes involved in sediment deposition and basin evolution. So far, we have seen how basins develop,

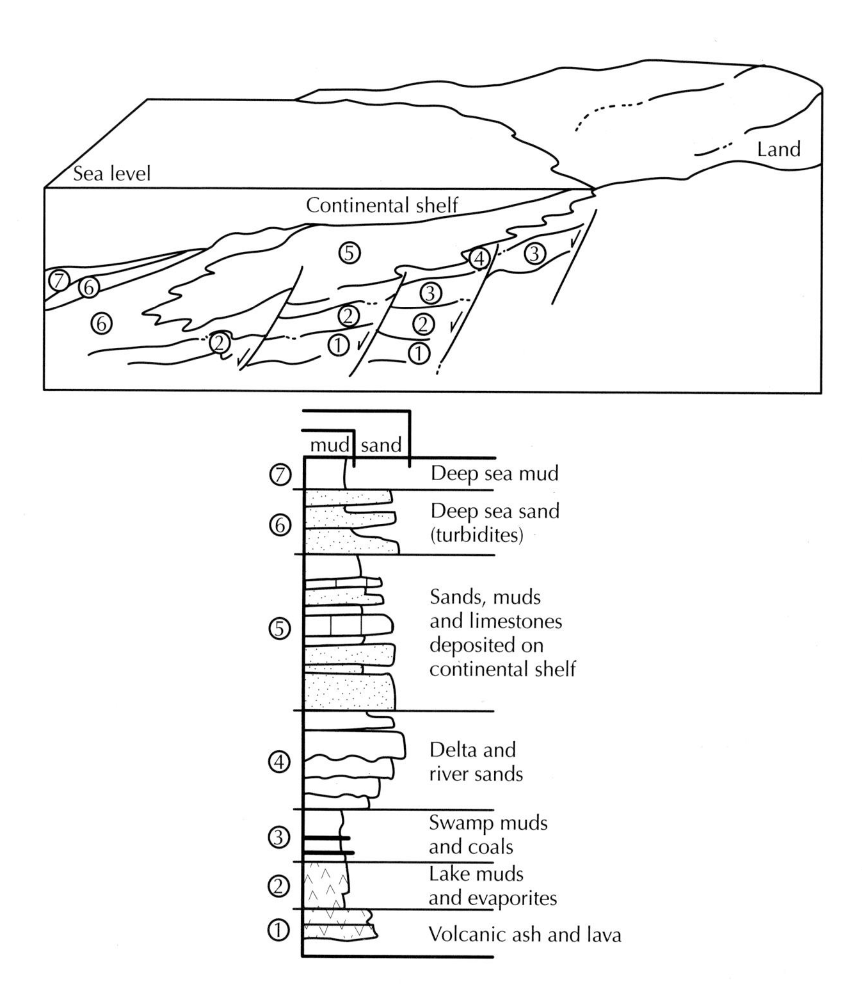

Diagram 5 Typical sequence of deposition at a developing continental passive margin. The sequence of deposits (1–7) records the rifting and subsidence of the margin, followed by flooding as it subsides beneath the sea.

how much time is involved in their development and the basic types of sediment which can accumulate in them. In order to interpret the origin of sediment deposits and fit it into a broader picture of the evolution of the sequences of deposits filling sedimentary basins, we now need to consider the processes involved in sediment deposition.

3

Evidence of ancient environments

Sedimentary rocks record the environments in which they were deposited – perhaps a warm, shallow sea or a desert. By looking at the sequence of rocks built up in a place over a period of time, we can see how the environment has changed there over the millennia and work out details of Earth history. However, one particular rock type is rarely indicative of one unique environment. Mudstones, for example, can form in shallow lakes or in the deep sea; sandstones may form in deserts or on the beach so we cannot jump from the observation of simple rock types to instant identification of depositional environments.

Instead, we have to look for clues to indicate the processes operating during transport and deposition of the sediments. They may include fossils preserved in the sediments, traces of plant roots or evidence of ripples on the sediment surface caused by wind or water. By building up a picture of the processes at work, we can narrow down the possible range of environments in which the sediments could have been deposited. Once this has been done for a sequence of sediments, it is usually possible to be confident about its origin.

Short term processes

Processes that operate on geologically short time scales range from almost instantaneous events (e.g. landslides and floods) to longer term events (such as the growth of a coral reef) over the course of a few thousand years. These are the familiar events which we witness at the scale of human lives. It is fairly simple for the geologist to observe how they operate at present and how they affect sediment deposition in real life settings.

Obviously, a vast range of processes take place on the Earth's surface and a full description would be very long and tedious. Instead, let us build a general picture of the kinds of events of which geologists need to be aware when interpreting the origin of most sedimentary rocks. They are summarised below, starting with the creation of fresh sediment in upland areas by the erosion of older rocks and following the journey of transported sediments down to the sea. We can then think about the creation and deposition of indigenous sediments.

The main factors at work in upland areas are chemical and physical weathering. At any particular site, the style of weathering depends on the type of rocks exposed and the annual range of temperature and humidity. Because water expands when it freezes, rocks are gradually broken up by ice forming within cracks in the rock. Areas where the temperature varies from below to above freezing point are generally a source of abundant sediment. In cold upland areas, mountain glaciers – "rivers" of ice creeping slowly down hill under their own weight – transport large amounts of fragmented sediment of this type (photograph 5). During the Ice Ages, such glaciers spread into lowland areas and transported piles of sediment which have blanketed the more northerly and southerly continents. The deposits are generally poorly-sorted jumbles of boulders mixed with fine "rock flour", scraped from underlying rocks by boulders trapped at the base of a glacier as it moved. Repeated freezing and thawing is also characteristic of desert uplands where the lack of cloud cover often results in freezing nights followed by intensely hot days. Again, piles of fragmented rock are formed but these are not transported away by glaciers.

Mechanical erosion obviously occurs throughout the transportation of sediments, whether by tumbling pebbles along a river bed, scouring rocks by

Photo 5 The Umasi La glacier flowing northwards off the High Himalayas, northern India.

wind-blown sand, or by coastal erosion by waves and tidal curents. Generally, rock particles get smaller and more rounded as they are transported, which is a useful guide to the distance sediment has travelled before deposition. The surfaces of sand grains record the impacts and abrasions they have experienced since they were first eroded. High-magnification study of quartz sand grains can sometimes show whether they were transported by the wind (causing "frosted" surfaces) or glaciers (causing fractured surfaces resembling broken glass), but this is generally unreliable because sand grains eroded from pre-existing sedimentary rocks can show surface textures inherited from earlier cycles of erosion and deposition.

Chemical weathering occurs wherever rocks are exposed to the atmosphere but is most rapid in warm, humid environments. Many of the minerals in rocks form at high temperatures and pressures; they are not stable in surface conditions, breaking down to form clay minerals which hold water in their crystal structure and are essential for encouraging plant growth. Very fine-grained fragments of rock – known as *silt* when smaller than 0.06 mm in diameter – mix with clay minerals to make mud, vast quantities of which accumulate during weathering of rocks in humid environments together with some contribution from volcanic ash deposits (which themselves break down to form clay minerals).

Carbon dioxide from the atmosphere, dissolving in water to produce carbonic acid, makes rain acidic enough to etch limestone rocks. Stronger acids form below plant-rich soils (partly due to the addition of carbon dioxide formed by the plants to the water percolating through the soil, and partly due to the formation of organic acids). In humid environments, these acids can over time dissolve vast quantities of limestone, creating fissures and caverns.

Dissolved limestone is carried by the ground-water and is later precipitated underground as stalactites, stalagmites and other cave formations. In places such as the Mediterranean and the monsoonal areas of southern Asia, where the climate varies from wetter to drier, limited dissolution of limestone fragments in soils during wetter phases is followed by evaporation and drying of the soils. This results in precipitation of the dissolved limestone as a layer of cement a few feet under the soil surface, called a *calcrete layer.*

Having created sediment from pre-existing rocks by weathering, it becomes possible to transport and deposit sedimentary rocks. The main transporting agents are gravity, wind and water.

Rock avalanches – landslides – occur in a range of upland settings when unstable slopes fail during storms or earthquakes (photograph 6). Steep scree slopes, on which rock-slides are common, pass downhill into gentler slopes where transport is more episodic, so movement of sand and gravel in these areas takes place mainly during rare, large landslides. Mainly because

Photo 6 Landslide deposit in the Indian Himalayas.

of gravitational instability, the sediments avalanche downhill and flow as a jumble of dry rocks; the smaller fragments tend to settle downwards during flow, leaving the largest rocks on top. You can see this effect if you shake a bowl of muesli – the nuts and raisins concentrate at the surface as smaller grains and flour settles to the bottom.

If a high proportion of wet mud is mixed in with the larger sediment grains, the whole muddy mixture slides down-slope under gravity with the larger rock fragments floating on the surface. This type of flow is common on the slopes of volcanoes where abundant mud has come from the weathering of volcanic ash; it also occurs in other areas with a significant slope where mud is present. Mud-free ("sandy") debris-flows are also known, but are much less common. Having reached a flatter area, the flows eventually stop and "freeze" to form a chaotic mixture of debris.

Gravity-flows are also possible in areas of unstable submerged slopes such as the steep fronts of river deltas building out into the sea. There the flows comprise a mixture of loose sediment and water, forming a dense cloud of turbulent debris which flows down-slope to be deposited in deeper water. The larger rock fragments tend to settle to the base of the flow with the finest sand and mud settling out more slowly as the flow loses power. This results in a graded deposit, with the finest grains on top.

Wind-driven flows are common only in deserts and other dry areas.

The wind cannot drive pebbles, which are instead blasted by sand storms to give them a characteristic faceted shape and left up-wind forming stony deserts. Sand dunes of various shapes are built up on the down-wind side from sand dropped by the wind. Sand avalanches down the (down-wind) slip-face of such dunes, forming thin (mm-scale) layers with the largest sand grains on top, just as in the formation of landslide deposits. Sloping layers of sand accumulate against these dunes, preserving the shape of the dunes unless they are destroyed by later erosion. The movement of sand bars in rivers or the shallow sea can yield similar sand dunes under water (photograph 7). The lack of pebbles and the presence of avalanche layers in wind-blown dunes helps to distinguish them from subaqueous dunes.

Running water is the most common transport agent on land. Powerful floods are able to transport large pebbles and small boulders, although such coarse debris is generally transported in gravity-driven flows as described above. Most rivers carry a mixture of mud, sand and pebbles. The mud is carried in suspension as a cloudy, turbulent mixture with water and settles out only when the river loses power. Sand and gravel are carried along the bed of the river, forming shifting dunes that move downstream during exceptionally powerful flow.

There are many types of river, ranging from fast, shallow, pebbly streams pouring from melting glaciers to sluggish, meandering channels

Photo 7 Inclined layers preserve the shapes of sand dunes which formed in a river feeding a sandy delta over 300 million years ago in central Scotland.

flowing through swampy lowlands. Decades of effort have gone into categorising the different types and establishing "typical" sequences of sediment deposited by them. It seems now that the most effective method of interpreting ancient river deposits is to look at the small scale pattern of sediments present, deduce the processes acting in the ancient river and then compare them with those known to operate in modern rivers. The kinds of sand bars present within the river channel, the stability of the river banks, the shape of the channels in plan view and many other features all need to be identified.

Some rivers have a wide, branching maze of shallow channels separated by sand bars which move only during flood conditions; no permanent river banks are developed and mud transported through the river system is rarely deposited. Such "braided" river systems deposit sheets of gravelly sand, comprising nested stacks of dunes which accumulate against the edges of the channels and bars (photograph 8). Other rivers have more permanent, stable channels with muddy banks often fixed in place by vegetation. The channels tend slowly to migrate sideways, leaving a characteristic deposit of inclined beds of mud and sand that build up against the shallow slope of the river bank (photograph 9). The banks are occasionally broken during floods, allowing sediment to escape from the channels and deposit sheets of sand and mud across the low-lying "flood-plain". As they leave the channels, the

Photo 8 Sheet of gravelly sandstone deposited in stacked, braided channels. Pakistan.

Photo 9 Inclined layers of mud and sand mark the edges of migrating channels in the thick bed above the geologist. Central Turkey.

floods lose power so that coarser sand is deposited close to the channels. Sandy deposits of this kind help to build up the river banks as "levees". The flood deposits tend to form thin sheets of sand and mud with the coarsest grains near the bottom, and commonly preserve ripples formed by the flood waters. After the flood has receded, the sheets of sediment are often colonised by plants so such deposits may preserve fossilised plant roots. Major floods can also establish a new channel, leaving the old one to silt up as the main flow of the river follows the new channel; it is often possible to recognise abandoned channels in the sediments deposited by this kind of river.

When rivers reach a standing body of water – a lake or the sea – they lose the power to transport most of their sediment load and deposit it in a delta. Mud tends to be carried furthest so that sand and gravel concentrate in the delta itself. Sand bars are deposited, choking the river which divides into a network of divergent "distributary" channels. If the coast is affected by strong wave or tidal forces, these bars are shaped into characteristic patterns. Wave-dominated deltas have sand bars spread out parallel to the coast as storm beaches. These also occur in areas with no major deltas, where sand is concentrated at the shoreline to form beaches after transport in numerous small rivers. Tidal deltas have sand bars and channels elongated at right angles to the coast. The smaller-scale features of the deposits record evidence of the dominant forces shaping the delta – wave ripples, together with piles of wave-washed pebbles and shells record wave influence, whereas tidal currents deposit alternating layers of mud and sand as the tide ebbs and flows. Where neither waves nor tides have much influence (as in lakes and sheltered seas), the delta builds out by deposition directly from river mouths, producing a "bird's-foot" pattern as in the modern Mississippi delta. Deltas are home to a profusion of plant and animal life so that deltaic sediments tend to show abundant evidence of burrowing by worms and other invertebrate animals; plant root fossils are common.

Lake deposits can sometimes be recognised by the kinds of fossil animals, generally different from sea creatures, which are preserved in them. They also show limited evidence of waves and storms, and their sandy beach deposits pass in a short distance out into quiet-water muds. Lakes tend to dry up periodically, exposing the muddy shallows which crack as they dry before being buried by renewed deposition as the lake fills again; sometimes they preserve the footprints of animals which passed that way in times long gone. Such deposits are unlikely to be preserved in the higher energy conditions of a sea coast.

Most river-borne sediment is deposited in the sea close to the shoreline, but rarely stays put for long because of wave action, tidal currents and storms. Coastlines facing the open ocean might be affected by hurricanes which can generate waves affecting the sea-floor to depths of more than 100 m below sea-level. The sea-floor sediments pile up onto the coast as storm beaches and accumulate offshore into wave-formed sand bars; the shells of sea creatures are mixed in with the sand. During such storms, finer-grained sand and mud are typically carried into deeper water as turbulent mixtures of sand, mud and water similar to the gravity-driven flows we met earlier. Sheltered shorelines experiencing less storm influence show smaller scale deposits produced by smaller waves. A great variety of marine life occurs in shallow seas so fossil shells and animal burrows are very common in shallow marine deposits. Their study yields detailed information on the environments in which the enclosing sediments were deposited, although after death the fossils may have been transported by submarine currents from where they once lived.

Deeper sea floors, beyond the influence of storm waves, tend to be muddy, with occasional deposits of sand flowing offshore during major storm events. The flows become less common further from the shoreline where quiet-water muds predominate. In areas with local submarine slopes, perhaps close to the edges of deltas or along fault lines where the sea-floor is being deformed by plate tectonic activity, sediments may be transported down slope by gravity. The flows sculpt the muddy sea-floor into streamlined hollows which fill with the new coarser-grained deposits to form characteristic teardrop-shaped "flute casts", with the narrow end pointing down-stream. The new deposits accumulate rapidly, in some cases forming sheets of loosely-compacted sand and mud several metres thick in only a few hours. Water escapes as the sediments settle, forming water-escape patterns in the sediment which provide good evidence of their rapid deposition.

We saw earlier that indigenous sediments generally accumulate in areas which transported sediments fail to reach. For example, raised areas on river flood-plains tend to be vegetated, swampy environments in which abundant dead plant material collects, turning to peat as it decays and likely to form coal if it is buried. So, in a sense, coal forms by default in areas of the floodplain where nothing else is being deposited. Swampy deltas, between the active distributary channels, are similar sites of coal formation.

Sea and lake water contains carbonates, chlorides, sulphates and other dissolved salts derived from dissolution of rock particles. Because they are fairly well mixed, the salt content of the oceans is much the same everywhere (about 3%). Lakes contain a mixture of salts derived from the local area, which may include volcanic minerals leached from rocks or supplied by springs. If a sea or lake dries up, the dissolved minerals are precipitated, the least soluble minerals coming out of solution first. Rock pools by the sea often have white rims of lime mud (calcium carbonate) and gypsum (calcium sulphate), two of the least soluble and most common minerals in sea water. Salt (sodium chloride) is precipitated only when the sea water is almost completely evaporated and in strongly evaporated seas, scums of salt crystals float as rafts which build up at the surface before sinking to the sea floor. Provided there is a periodic, limited resupply of sea water into the evaporating basin, thick deposits of salt plus other minerals (*evaporites*) can develop in this way without the sea drying out completely.

Precipitation of lime mud also occurs at springs and in caverns, forming crusts of tufa (at springs), stalagmites and stalactites (on cave floors and ceilings) and other similar deposits. This is mainly due to the release of carbon dioxide from solution as the water reaches the open air, making the water less acidic, so it can carry less calcium in solution and a proportion is precipitated. Anyone living in an area with "hard" tap water will be familiar with this effect. Some animals and plants make use of this phenomenon to precipitate hard, limestone shells to support and protect their bodies, something they have done for over 500 million years. Strangely enough, calcium ions are actually toxic to most organisms, suggesting that calcium carbonate was originally deposited by primitive creatures as a waste-disposal mechanism which resulted in shell formation as a useful by-product. The shapes of the shells are characteristic of the animal species that formed them and they help us to trace the evolution of life. The shape of a shell is to a great extent governed by the use to which it was put, thus providing evidence of the way of life and habitat of the organism that used it. Study of modern organisms has identified shell forms which are characteristic of creatures living in specific environments, so we can deduce a great deal about the way of life of extinct creatures from their fossil shells and skeletons. If we can be sure that the creature lived at or close to the place where its shell was found, we can learn a lot about the depositional environment. For example, thick-shelled sea creatures typically live in areas affected by strong waves – it is not worth the effort to build a thick shell unless you need it for survival in rough waters. Some shells are elongated, adapted for burrowing into the sea-floor for food or protection. Others are thin and streamlined, used by creatures swimming near the sea surface. All contribute to an understanding of conditions during their lifetimes.

Of course, most shells are not preserved in the place where their owners lived. Enclosed shells may fill with gases during decay and float around in the sea before being punctured and settling. Others are moved by storms and tides along with sand and mud. Microscopic sea-shells may even wash up on a beach and be blown inland by strong winds. But generally, if a large shell occurs in a deposit composed of much smaller, lighter clay and silt grains it is likely to have dropped in after the nearby death of its maker. And, as with coal and salt deposits, in the absence of transported sediment such indigenous deposits can become the dominant rock-forming component. In the deep sea, microscopic shells formed by the surface-dwelling plankton rain down constantly onto the sea floor. Some are dissolved on the way to the bottom and others are dissolved on the sea floor, but sometimes this slow "snow" of shells accumulates to form limestone: chalk is a good example of this kind of deposit. Shallow marine limestones, formed from the shells of nearshore creatures, often contain some mix of mud and sand in addition to the shell material. They form part of a continuous range of shelly mudstones, shelly sandstones, sandy limestones and muddy limestones.

The purest nearshore limestones form in areas where little sand and mud comes from the shoreline, for example off the windward coasts of flat-lying deserts or vegetated tropical islands. In such conditions, shelled organisms have evolved to construct stable frameworks, raising themselves into the light and warmth of the shallow sea. In modern, tropical seas these "reefs" are typically built by colonies of corals, but a similar role is played today in colder temperate waters by bryozoa and algae. Other organisms built reefs in the past; for example, most reefs created during the Cretaceous period, when dinosaurs walked the land, were composed of colonial animals similar to tall oysters. Reefs provide barriers behind which protected lagoons are populated by other shelled organisms, so forming their own characteristic assemblage of shells adapted to their preferred habitat. Reefs can build upwards if sea level rises or the land subsides, creating steep slopes on the seaward side down which gravity flows of shelly debris and fragments of reef rock occur.

Longer-term processes

With an understanding of the short-term processes observed in present-day environments, it is often possible to deduce the events responsible for deposition of single units ("beds") of sedimentary rock. They are typically a few cm thick and consist of sediments deposited by a single event, such as a flood or a storm. Fossils, burrows, roots and ripples and other clues help to establish the environment in which the deduced processes took effect. The next step is to work out how the environment changed over time, resulting in the deposition of a succession of different beds of rock.

We know that younger deposits pile up on top of older ones. It is also fairly obvious that, in a continuous sequence of deposits, a gradual change up the pile from one environment to another records a change to a new environment that existed side-by-side with the old one, in a geographical sense (diagram 6). For example, if you find a pile of sediments with river deposits on top of desert deposits, the river must have been flowing through the desert and the sequence of rocks records the river migrating into the local

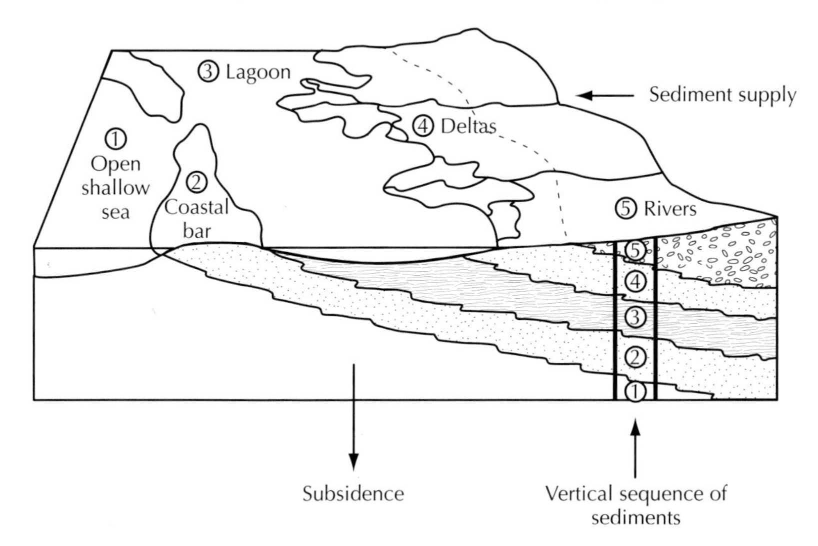

Diagram 6 Sketch showing why a vertical pile of different rock types in a continuous succession of deposits records deposition in adjacent settings.

area. This assumes, of course, that the sequence is continuous. An alternative possibility is that the river deposits are much younger than the desert deposits and were deposited after changes which are not recorded in the rock sequence (perhaps because of erosion after the desert sediments were deposited).

We cannot observe such longer-term processes in nature because of our relatively short life spans. We have to note sequences of sedimentary rock deposits, deduce the changing sequence of processes and environments which they record, and then ask what brought about the observed changes. There is a degree of theoretical interpretation, partly involving judgements of what is likely or unlikely to bring about changes in the environment. There are risks attached and theories change as new processes and environments are observed and described. Still, there are some well-established principles which can be applied to most rock sequences.

We can envisage the way changes in environmental conditions can affect sediment deposition by thinking about a simple set of environments and how it would respond to particular changes. Imagine a delta where a river flows into a small inland lake. Generally, sand and gravel are deposited in the river channel, finer sand is deposited in the delta as the river slows down on entering the lake and mud is deposited in the quieter waters of the lake in front of the delta. Now assume that the whole area is subsiding as part of a sedimentary basin. A vertical sequence of sediments is piled up, recording changes in the environments of deposition with time. We can look at vertical changes in the sediment types, interpreting them in terms of changing processes.

The simplest sequence to think about is one in which there is no change in conditions with time. River deposits pile up on top of older river deposits and the position of the delta remains fixed. For this to occur, there must be a balance between the subsidence of the basin, the amount of water in the lake and the supply of sediments from the river. Subsidence creates space for sediments to be deposited. The amount of water in the lake determines how much of this space is available to be filled by the sediments coming from the river and its delta. This potential sediment repository is called *accommodation space*. The balance between changing accommodation space and fluctuating sediment supply is the key to understanding sedimentary sequences.

Imagine now a gradual change from finer-grained, muddy lake sediments to coarser sand and gravels deposited in the river – a *coarsening-up* sequence. Clearly, the river must have built its delta out into the lake, a process known as *progradation*. This might be accomplished either by a reduction in accommodation space or an increase in sediment supply. The former means that either the volume of water in the lake fell, perhaps because of a change to a drier climate, or the sedimentary basin was subsiding more slowly, maybe as the consequence of a change in the plate tectonic forces that were causing it to subside. An increase in sediment supply could be due to a change in climate, causing the river to flow more powerfully and transport more sediment, or to a change in the slope down which the river flowed resulting from plate tectonics. Smaller scale, local increases in sediment supply might follow from diversion of the river channels to flow into new areas of the lake margin. In either case, the delta was forced to prograde out into the lake because of the changing balance between accommodation space and sediment supply.

These principles can be applied in reverse to explain changes from coarser river or delta sediments to muddy lake sediments. The delta must have been drowned by the lake expanding either because of an increase in accommodation space (maybe the lake level rose – the climate could have become wetter; perhaps the basin began subsiding more rapidly) or a decrease in sediment supply causing the delta to retreat.

Choosing between these various possibilities is not simple. One has to look at the sequence of sediments recorded at the same time in different areas to determine whether the changes were local or regional. Local changes are more likely to have come from modifications in the flow direction of the river channels, switching the position of the delta from place to place. Regional changes in climate or lake level may be reflected by the kinds of sediments deposited in the deeper parts of the lake, away from the input of sediment from the delta. Fossils, burrows, animal footprints, plant roots and mud-cracks in the lake sediments can all provide clues to changes in lake level which might explain changes in sedimentation at the delta. Changes in the composition of sand coming down the river can reveal the consequences of changes in sediment supply from upland areas, linked in turn to plate tectonic or climatic events.

The same thinking can be applied to any area of deposition. There is always a balance between sediment supply and accommodation space, and shifts between them can usually explain observed changes in the sequence of sedimentary deposits. For example, coral reefs expand and contract due to changes in the rate of reef growth (i.e. sediment supply) and changes in sea level (i.e. accommodation space). Progradation of reef margins (out-building) can be observed in limestone cliffs and interpreted in terms of falling sea-level or increasing reef productivity.

To understand changes in deposition recorded by sequences of sedimentary rock tens to hundreds of metres thick, we have to consider processes operating over tens of thousands to millions of years. On this time scale, changes in accommodation space occur because of (a) the creation and evolution of subsiding basins under the influence of plate tectonics, and (b) fluctuation in absolute sea-level due to the melting of polar ice caps and the growth of submarine mountains and volcanoes which displace sea water onto the land. Variations in sediment supply on this scale happen when land is uplifted to supply additional sediment by erosion and major climate changes alter the rate of sediment erosion from existing upland areas.

Documenting the evidence for all these effects requires careful and consistent recording of rock properties so that observations by different people at different times can be compared directly. This is the job of geologists and geophysicists in the field; in the next section we will get a sense of what that job involves.

Measuring rock properties

Most of my working life has been spent in the field, measuring rock properties and interpreting them in terms of the evolution of sedimentary environments. The main reason I wanted to work as a geologist was the attraction of getting out into the wilderness with a tent and rucksack, and finding out about ancient landscapes and the creatures that lived there. It turned out that although this romantic notion of field geology does form part of Earth science, there are many other sources of geological information – and many of them are just as fascinating. Modern technology, including getting information from deep boreholes and interpreting satellite images, has for many geologists replaced the rock-hammer and rucksack. But all the tools available to the Earth scientist have to be used to build a more complete and more accurate picture of Earth history.

Geological field-work

Basic geological field-work still plays a major part in the science and remains most people's idea of what geology is all about. Many office-based geologists pine for their student days when they worked at the outcrop and most of them appreciate a chance to go on an organised field trip. Interested amateurs can – and do – contribute vital new information from their own observations in the field, particularly by discovering new fossil species.

Novel field techniques and enlightened understanding of geological processes means that areas which have been studied in the past continue to yield new insights into Earth history. The many important scientific papers published every year from re-examination of outcrops in Europe and America show that you do not have to go far afield to find new information at outcrop. However, professional field geologists employed by international energy and mining companies tend to work in remote parts of the world where the chances of finding undiscovered resources are greater than in more familiar terrain. Doing your job in remote areas inevitably sets some limits on what you can observe, mainly because of the accessibility of outcrops and the time available for documenting their properties.

My own experience of geological fieldwork has been in Europe (mainly the UK and southern Spain), Russia, Azerbaijan, central and southern Turkey, Syria, Yemen, Oman, Pakistan, India and Nepal. Most has been on behalf of major oil companies who provided logistical support. Access to remote areas is generally possible by helicopter (unless the locals are given to shooting at them), four wheel drive vehicle, or more humble means (photographs 10a–c). In some parts of the world, armed guards are needed for security reasons. Good medical support is often essential, particularly in areas where snakes or disease-bearing insects are common. Sometimes it may not be of much help, as in western Pakistan where some species of sand fly carry an incurable, often fatal, wasting disease. But the risks should not be exaggerated: field-work is probably safer than walking the streets in many Western cities and in most countries the greatest danger is driving on the local highways!

Having reached an outcrop, the primary task is to document the

a)

Photo 10 Access to the outcrop: a) by camel, Pakistan; b) by horse and c) on foot, northern India. *(continued)*

b)

c)

Photo 10 *(continued)*

sequence of exposed rocks. Because of folding and warping of the rock layers after deposition, most rock layers are tilted at some angle from the horizontal; this tilt is called the *structural dip*. Detailed measurements of the dip and other aspects of the folding and disruption of rock layers is the task of the *structural geologist*; they record all aspects of the Earth's crustal deformations and they can work out how and why the structures formed. Such information links the general plate tectonic setting of a basin, generally already known from previous studies, to the development through time of the local structures. Local areas of increased subsidence, forming small sub-basins, can be defined, allowing predictions to be made about the pattern of sediment transport into the basin during its history. Detailed knowledge of the surface structure is also used to predict the subsurface structure and suggest, for example, where pockets of oil and gas may have accumulated.

Although I have tried my hand at structural geology, my main interest is in studying the sequences of sediments which have been exposed by warping of the crust and subsequent erosion of the tilted layers. By walking along the direction of structural dip, one encounters younger and younger layers of rock in what was once a vertical sequence (diagram 7). The thickness of the layers is measured using a tape or a pole marked with 10 cm intervals. Accurate thickness measurements are easier to make if the pole is fitted with a level which can be adjusted to the angle of dip. I generally use a detachable level which can be fitted to a broom-stick or similar staff after arriving in the field area. The total thickness of the sequence at a locality,

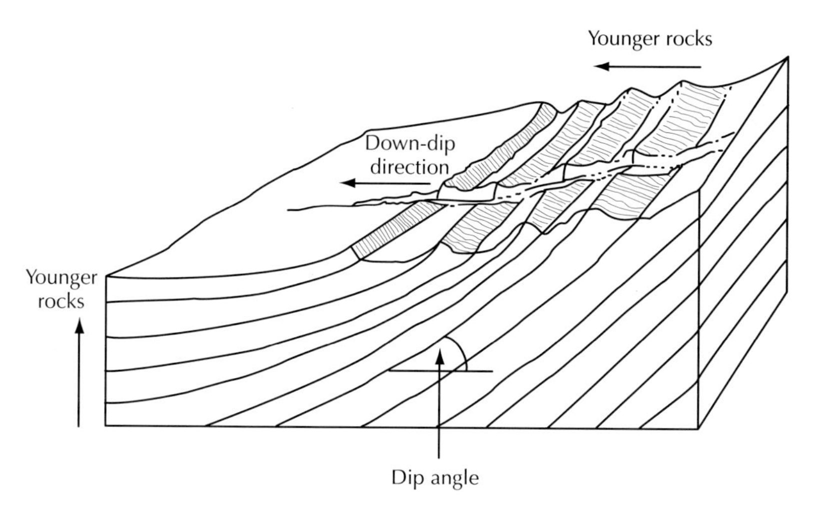

Diagram 7 Tilted layers of rock "dip" at an angle relative to horizontal. Progressively younger rocks are encountered when walking along the down-dip direction, as shown.

comprising exposed rocks and unexposed, covered sections, is an important measurement in a regional study of variations in sequences from place to place.

The sequence has to be divided up into manageable units for detailed description. The scale of units chosen depends on the amount of detail required and the time available for description. The main requirement is to define units which were deposited by a single type of event. In order to gather enough information for the interpretation of depositional environments, it is usually necessary to describe individual units a few tens of centimetres to a few metres thick, representing deposition from separate flow events. They may range from gravity-driven deposits 10 m thick originating in a single landslide event, to millimetre-scale layers of sand deposited by storms in a lake. Once sufficient detail has been recorded, groups of units can be lumped together to save time, perhaps by measuring the thickness of sets of gravity-deposits that thin or thicken upwards, or by measuring the entire thickness of a sequence of alternating millimetre-scale muds and sands.

Each unit is characterised in terms of rock type, grain size variations, colour, internal structures and fossil content. The rock type is classified according to standard categories of sandstone, mudstone and limestone, together with a few other, less common types (such as coal). For example, a sandstone composed entirely of quartz grains is called a *quartz arenite*; a limestone made up only of shell debris is a *bioclastic grainstone*. Out in the field, tools are necessarily simple: a hand lens (typically with 10× magnification), a knife (which will scratch limestone but which will itself be scratched by quartz), dilute hydrochloric acid (which causes calcium carbonate but not quartz to fizz) and possibly other specific tests; full description may have to wait until samples are back in the lab. Categories of common sedimentary rock types are shown in diagram 8. Note that the limestone categories are mainly based on recognising the proportion in the rock of *lime mud*, a very fine-grained calcium carbonate produced mainly from the breakdown of loosely-aggregated skeletons of some organisms such as calcareous algae. Lime mud is produced mainly in the sea and should not be confused with mud transported from the land, which is composed mainly of clay minerals and silt-sized rock fragments.

The grain size of transported sediments is categorised according to a scale which varies from clay and silt to various kinds of gravel. Grain size categories are defined by exact dimensions which are generally impossible to determine in the field. For example, a sandstone may be composed of a mixture of grains of various sizes. Generally, the approximate average grain size of a rock is determined by visual comparison with a standard scale

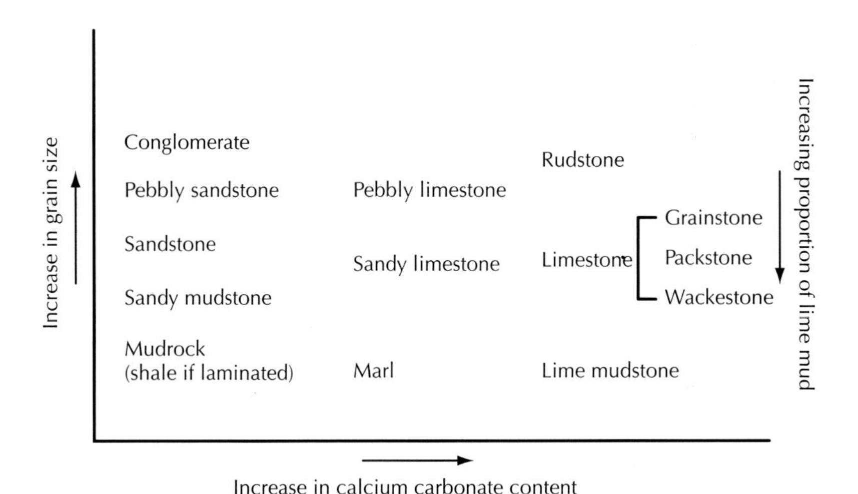

Diagram 8 Range of rock types, named according to their grain size and calcium carbonate content.

called a *grain-size comparator*, small enough to be carried in the field. It can also be used to estimate the degree of sorting of the grains into particular size categories (a well-sorted rock consisting of grains which are all about the same size), the roundness and shape of the larger grains, and the percentage of distinctive grains visible.

Rocks are not generally very colourful but their colour can give an important indication of composition. This is particularly true of mudstones which commonly contain significant amounts of very fine-grained iron minerals. Oxidation of the iron minerals, either during or after deposition, leads to a range of colours – black muds contain no oxidised iron, whereas grey, green, yellow, orange and red muds contain progressively more. Rock colours tend to be a subtle shade of grey, so that one soon runs out of variations on "greyish", "buff", "pale brown" and so on. Field geologists tend to describe rock colours in rather extreme terms, referring to greyish rocks as "green" or "purple" to emphasise their differences. A system of colour categories has been defined using a standard comparison chart. Colours are described by a letter and number code (so that pale yellow might be referred to as 10YR 6/4, and greenish-grey as 5G 5/1) but this is not readily understood by non-specialists. Probably the best way of describing rock colours is by photographing a typical example and describing significant departures from this norm.

Internal structures within sedimentary rocks are known as *sedimentary structures.* Whole books have been devoted to describing and illustrating their range and determining their origin. They are important because they give direct indications of the processes prevalent at the time of their deposition and shortly afterwards. The student geologist can gain a lot from reading such books, but learns most from seeing an experienced teacher demonstrate the identification of sedimentary structures at real outcrops. The main types of sedimentary structures which are valuable for determining depositional processes are:

1. *Contacts between rock units.* Gradational contacts record a slow or continuous change from one rock type to another, for example from sand to mud as transport capacity falls in a decelerating flow. Sharp contacts usually signify a break in deposition, of unknown duration. Irregular contacts may be due to the upper unit sinking into the lower one while the latter was still soft and wet. Cross-cutting contacts indicate erosion of the lower unit by currents before or during deposition of the upper. Lateral variations in unit thickness suggest erosional truncation of units or draping of units onto a pre-existing, uneven surface.
2. *Internal depositional structures.* Internal bedding, seen as variations in colour and/or grain size, has a variety of forms. Flat, millimetre-scale layering is called *lamination.* Muds split easily along lamination, in which case they are said to be *fissile.* Laminated muds are called *shales.* The presence of lamination in muds is due to the presence of microscopic layering which formed as the mud settled from suspension, and this shows that the mud has not been mixed by worms or other burrowing organisms, which in turn suggests that no oxygen was present in the water when the mud was deposited (see 3, below). The uneven surfaces of ripples may be preserved in the sediment as they are buried, forming *wavy-* or *cross-lamination,* the shape of the laminae depending on the shape of the original ripples. Wave ripples have wavelengths of up to about four times their crest height whereas ripples caused by flowing currents have longer wavelengths in comparison to their height. Ripples may also be preserved on the surfaces of sediment layers (photograph 11).

   Larger sand dunes, formed by flowing wind or water, create *cross-bedding* within the sediment. The direction of current flow can be determined from the shape of the dunes which generally are steeper on the down-current side. Many dunes are crescent-shaped in plan view so that their flow direction can only be measured accurately from three-dimensional exposures. Very rapid, shallow flows often produce flat, thin beds of sand with characteristic streaks called *primary current lineation,* showing the flow orientation. This can be seen on the surfaces of flag-

Photo 11 Large-scale wave ripples on the surface of a shallow marine sandstone bed, Pakistan.

stones used in pavements in many older cities. Cross-bedding is also common in gravels, recording the migration of gravelly bars in strong currents, and in transported limestones. Upward variations in grain size through the unit can be found in graded beds with the largest grains at the bottom, or reverse-graded beds where they lie at the top. As we saw earlier, the degree of sorting of the sediment, particularly the presence of mud within the sandy or gravelly units and of anomalously large grains (such as pebbles) "floating" in a matrix of smaller grains (such as mud) can help to distinguish various types of gravity-driven deposits.

3. *Post-depositional structures* are produced mainly either by deformation of the sediments shortly after deposition, or modification by organisms. This includes deformation of beach sediments by the impact of storm waves, of sand dunes by waves and currents, and post-depositional features such as mud-cracks, rain prints, footprints, plant roots and fossil lightning-strikes. Rapidly-deposited wet sediments collapse and settle into a denser packing by losing most of the contained water soon after deposition, producing *water-escape structures*. One can see folds and breaks within the sediment, "dish and pillar" structures, and miniature sand volcanoes which erupted escaping water at the sediment surface. Sand entrained in escaping water may be carried upwards through

> overlying sediments during the de-watering of buried deposits, forming cross-cutting sandstone "dykes" of various sizes. Burrowing organisms produce a variety of tubular structures along or across sediment layers, recording the life-style of the organism. Some, adapted to specific environments, produce characteristic burrow patterns. Complex or meandering trails tend to be produced by worms feeding at or below the surface of the sediment in the deep sea while mud-lined burrows in sand argue for a tidal environment, because the mud linings help to stop the burrow collapsing when the tide goes out. Intense burrowing may obliterate former lamination, producing a featureless or mottled appearance. Preserved lamination in mud deposits generally points to insufficient oxygen for burrowing organisms to churn through the sediment. This has a particular significance for oil exploration because crude oil is generated from mud rich in organic matter which is generally deposited and preserved in poorly-oxygenated conditions hostile to burrowing creatures.

Other post-depositional features record the transformation of the deposited sediment into rock and its history of burial, deformation and uplift. These features include lumps or "nodules" of cement (crystallised from percolating waters) which may grow within the rock, colour banding due to oxidation or reduction of rock minerals by percolating water or other fluids (which can be mistaken for original bedding layers), and small scale folds and crinkles which may be mistaken for sedimentary structures such as ripples. Compaction during burial can squash original features, making them more difficult to recognise, and under high enough pressure this may even cause dissolution of the rock (especially limestone).

Fossils are particularly important in determining the depositional environment of rocks in which they are found (especially if there is reason to believe they lived close to their burial place) and may also be helpful in deciding the age of a rock.

The details of rock type, grain size, colour, sedimentary structures and fossil content are usually recorded in the form of a "log" in which details are filled in under particular headings for each rock unit; this helps to avoid forgetting aspects of the description. The exact style of the log depends on the preferences and objectives of the person making the description. In practice, it is best to have two geologists working together, one defining and measuring the thickness of rock units, the other recording the details on the log. Both can observe and describe details of colour and sedimentary structures and look for fossils. In this way, important details are less likely to be overlooked and the description will be less biased by one person's preconceived expectations of what is likely to be observed (since it is often the case – and not only in geology! – that you tend to find what you are looking for). Photographing key observations is essential: it helps to prove the existence of what you are describing and greatly facilitates re-interpreting observations in the light of new ideas.

"A picture", they say, "is worth a thousand words"; an actual sample is worth a lot more than that. Indeed, the economic value of a rock sample is proportional to the effort expended in getting it – ones from remote parts of the world, deep oil wells and other planets are very precious stones. Choosing and cataloguing them clearly requires some care. Sampling strategies are mainly governed by what you want the samples for. Representative samples are generally collected from typical rock types encountered in a sequence together with any unusual oddities. Fist-sized specimens are required for determining such standard properties as the capacity of rocks to hold fluids (*porosity*) and to allow fluids to flow through them (*permeability*) – both key properties in the oil, gas and water industries. Larger chunks are needed to look for rare components like apatite grains (see Chapter 4). Samples are commonly marked with a pen to show their orientation, usually the way up, to help in later description and interpretation of sedimentary structures; they can be cut into thin sections (about 0.03 mm thick) for examination under the microscope. The precise orientation of the sample relative to the Earth's magnetic field has to be recorded for samples taken for palaeomagnetic analysis (see Chapter 5). Oil companies often need to know how much hydrocarbon could be generated from muddy sediments if they were buried and heated sufficiently – to find out, fresh rock samples are needed for laboratory analysis, which may mean extensive excavation or even drilling of the weathered outcrop. Large fossils may need to be dug out of the rock, even using explosives or rock drills, and transported in a plaster of Paris coating.

Seismic surveys

Structural geologists can often make reliable predictions of the shape and developmental history of subsurface structures on the basis of surface observations. Better still, the structures can be imaged directly by *seismic surveys.* These cost ten times as much as a field survey by a structural geologist so they are typically conducted by energy or mining companies to survey particularly prospective areas. Seismic surveying is a kind of echo-location technique. Shock waves are created at the surface and travel down through the underground rocks. Some rocks carry shock waves faster than others – muds, for example, are poor transmitters of shock waves, whereas hard limestones "ring" much better; such rocks are said to have different acoustic properties. Wherever there is a contrast in acoustic properties, for example at the top of a hard limestone layer buried under mud, some of the shock-wave energy is reflected back towards the surface. An array of geophones (microphones designed to pick up the returning sound waves) at the surface listens to the echoes, recording their arrival time after the initial shock wave was set off. Analysis of the arrival times measured at different geophones allows the geophysicist to work out the depth to the reflector and its angle of dip. The properties of the reflected wave also carry some information about the type of reflector which created them. Subtle changes in the travel time of the shock wave through the rock can be used to infer changes in the rock type along layers; they suggest the locations of sandier and muddier areas on the buried surface as well as changes in the fluid content (perhaps gas instead of water) of the underlying rocks. In modern oil exploration and production, dense arrays of geophones are used to map out the subsurface structure in three dimensions (*3-D seismic*) and to trace changes in fluid content as oil and gas are extracted from a producing field and replaced by water (*4-D seismic* – the fourth dimension being time).

Remote sensing

Remote sensing is the observation of the properties of rocks (or other objects) at some distance from the observer. The term could be applied to seismic surveying but normally refers to airborne or satellite observations of the surface and subsurface. Observations like those cover large areas of territory and are typically employed at the early stages of oil, gas or mineral exploration to delineate areas of interest to be surveyed in more detail on the ground.

In virgin territory, remote sensing is used to detect thick piles of sediment filling sedimentary basins and to map the shapes of the basins in three dimensions. The techniques include:

- magnetic surveying which shows the position of buried magnetic rocks such as volcanic and igneous masses;
- gravity surveying, contrasting the weak gravitational pull of the less compact sediment in a basin with the stronger pull of dense, older rock surrounding the basin;
- airborne surveys to detect hydrocarbons leaking from underground oil and gas fields at sea, provided the slicks created by sea-floor oil seeps can be distinguished from man-made spills;
- buried detectors to sense gas seeps on land, although the presence of leaking gas does not necessarily imply a commercial field at depth.

The contrasting colours and weathering patterns of different rock types makes them fairly easy to distinguish from the air in sparsely-vegetated terrain, allowing surface maps to be made from air photographs. Pairs of photographs are taken of each section of terrain, at slightly different angles and from slightly different perspectives. The effect is the same as using two eyes to view a scene – a three dimensional image. Stereo imaging of air photographs shows up terrain topography and the orientation of tilted layers of rock exposed at the surface. Accurate measurements of structural dip and observations of rock types, made later in the field, refine the initial interpretations. In areas with flat-lying rock layers, it is even possible to see the traces of ancient river beds and related features which help to define past sedimentary environments.

Satellite images cover larger areas than air photographs but in less detail. They help to define large-scale structures and to detect subtle changes across large areas which may not be appreciated on the ground. Satellite-borne detectors can also survey the ground at wavelengths beyond the visible range, partly to penetrate cloud or vegetation cover and partly to gather additional information on surface rock properties. Just as different rock types display a range of colours by differentially reflecting visible light, they also do so in the infra-red and ultra-violet parts of the spectrum to generate "colours" visible only on photographic plates of special instruments. Rock properties revealed by such imaging are very helpful in large-scale mapping but do not include the detailed information required to reconstruct past depositional environments.

Drilling, coring and logging wells

The best way to find out what lies underground is to drill down, pull rock up to the surface and take a look. The cost involved in drilling a well – maybe ten times that of a seismic survey, or a hundred times more than a geological field survey – makes this an expensive luxury and the deductive exploration methods we have already surveyed represent a cheaper compromise. The expense of drilling must be justified by extracting as much information as possible from the rocks penetrated by the drill-bit.

Oil and gas are generally found at depths of 1,000–5,000 m below the surface, where temperatures may exceed 150°C and pressures are measured in kilobars (thousands of times greater than atmospheric pressure). The primary consideration of the driller is to keep the drill-bit lubricated and turning, and to maintain enough pressure in the hole to contain any pockets of oil or gas encountered. As the drill drives into the hole, a stream of mud – seeded with dense materials to increase its weight – is circulated by a pump; the mud keeps the bit lubricated and contains fluids released from the rocks at depth. A series of valves at the surface can be used to close off the hole and shut in the fluids if all else fails: this is a *blow-out preventer*; its use means that, unless something goes horribly wrong, the spurting "gushers" of Hollywood fame no longer announce an oil discovery.

The mud flows down to the bit through the drilling pipe to which the bit is attached. Additional lengths of pipe (each about 10 m long) are added as the bit penetrates deeper into the rock. Adding pipe involves pulling the bit up the hole far enough to unscrew the top of the pipe from the rotational drilling motor, so three 10 m sections are generally added at each step to save time. Mud flows out around the rotating bit and then back to the surface between the pipe and the walls of the hole, taking with it released fluids and fragments of rock cut by the bit.

At the surface, the mud is analysed to determine the type of fluids coming up from depth, so providing an early indication of gas or oil discoveries; it is then sieved for re-use and circulated back down the pipe to the bit. The record of the fluids contained in the mud during drilling is called a *mud log*. The main purpose is to detect hydrocarbons released from the rocks being drilled. The *cuttings*, rock fragments sieved out of the mud, are collected, washed and examined by a geologist. Changes in the type of rock emerging from the hole normally indicate the penetration of a new layer although the depth to the layer is not known for certain at this stage because of the possible variation in travel time of cuttings from the bit to the surface. Moreover, fragments of rock units which have already been penetrated at shallower depths may cave in from the sides of the hole so giving a false impression of what is currently being penetrated by the bit.

Caving in of the wellbore is particularly common in mudstones which swell and flake off as the drilling mud washes over them, while rock salt and other evaporite rocks may dissolve as they are drilled. In some cases these problems can be alleviated by using drilling mud mixed with oil rather than water but it is expensive, can be environmentally damaging and may hamper the subsequent evaluation of rock and fluid properties. At some stage, the wellbore is shored up by cementing in place a steel jacket or *casing* which lines the hole and prevents further caving. To do this, the drill bit is pulled out of the hole and the casing is inserted. A narrower drill bit is then fitted which can be lowered through the cased hole to drill the next section of the well, after which further casing is set. Thus, the larger 16–20 inch (40–50 cm) diameter bits used near the surface are reduced in stages down to about 6 inches (15 cm).

Representative samples brought to the surface from each section of the well are characterised in much the same way as rocks at outcrop although the small size of the cuttings makes most sedimentary structures impossible to identify. As samples may also be badly deformed by the bit, only the most compact and hardest parts of the rock generally survive. The identification of microscopic fossils ("microfossils"), indicating the age of the rocks being drilled, is one of the principal uses of these cuttings. At the well site in addition the mud log records are examined and some samples are routinely immersed in an organic solvent and illuminated with ultraviolet light: hydrocarbons leaching out of the sample fluoresce, showing that the drill has reached an oil- or gas-bearing level. Sudden changes in the rate of penetration of the rock by the drill, and of the fluid pressures in the well, may indicate that a porous, pressurised layer has been encountered.

Although cuttings are plentiful and easy to obtain, we have seen that they may not be truly representative of fragile rock types because of damage or caving in. Once the drill has reached a depth where hydrocarbons are expected, more reliable methods are required: direct sampling of the rocks themselves and remote sensing.

Rocks are sampled directly by coring at a known depth and recovering the cores to the surface:

- *Sidewall cores* are cut by a small metal cylinder which is forced into the side of the hole at a selected depth, before the casing is set. The cylinder is withdrawn, hopefully containing a rock sample, and sealed by remote control before being winched to the surface. Samples are often fractured by the coring action but their depth is known and they are particularly useful for age-dating using microfossil analysis (see Chapter 5).
- *Full-bore cores* can be taken only from the rock as it is drilled. The whole of the pipe is pulled out of the hole and the bit is replaced by a special core cutter – a hollow metal tube with industrial diamonds or other abrasives fitted at one end. The corer bores 30 m or so down into the undrilled rock, cutting a cylindrical core of rock which is trapped inside the corer by a spring-loaded mechanism. The core is then pulled out of the hole, packaged in 10 m-long barrels, and shipped off to a laboratory for description.

In view of the expense of delay to drilling during coring, full-bore cores are cut by energy companies only when absolutely necessary to understanding how best to recover hydrocarbons. That is frustrating for the geologist who wants to understand the overall picture of basin development – cores cut despite the wishes of the drilling engineer and the accountant are treated with proper reverence. However, government-sponsored research drilling often includes coring of almost the entire drilled section, using specially-adapted drill bits and drilling rigs (and cooperative drilling engineers!). Thousands of metres of core have been recovered from both onshore and offshore wells by such agencies, providing an invaluable research resource.

Once in the laboratory, cores are usually sliced longitudinally into two or three slabs which are shared between the operating oil company drilling the well, the government in whose jurisdiction the well was drilled and possibly other parties. The flat face of the cut slab gives a perfect, clean "outcrop", superior in quality (if not in size) to most surface outcrops seen in the field. The slab is generally cleaned and made easier to describe by soaking in water, because many sedimentary structures, such as burrows, show up better when the rock is wet; descriptive logs and photographs are produced. The detail of the logs, and any sampling strategy, depends on the objectives of the description. The original orientation of the core has to be known if full use is to be made of any directional structures such as current ripples recording the direction of sediment transport. Some cores are marked with grooves as they are cut in order to show their original orientation.

Direct sampling of rocks at depth is expensive, and it has to be planned in advance; rocks already drilled cannot be cored. Various remote sensing techniques have been developed to explore rock properties after drilling – they provide most of the well information available to the geologist. The main one is the *wireline log*, a record of rock properties measured by a tool lowered down the well, after the bit has been withdrawn, and then slowly winched back to the surface. As the tool travels up the well, signals are transmitted to the surface along the suspending cable. The measurements range from a simple gauge of the hole diameter to high-resolution visual images of the borehole wall. Many of the measurements cannot be made through casing so wireline logging is generally conducted before a drilled section is cased off.

The most common measurement (which can be made through casing) is of the rock's natural radioactivity. All rocks have a low level of natural radioactivity originating from various amounts of the radioactive elements uranium, potassium and thorium, the last being the most common. Mudstone generally contains much higher concentrations of thorium than sandstone or limestone. Uranium occurs in sedimentary rocks, mainly in association with preserved organic matter (more common in muds than in sandstones or limestones) which adsorbs dissolved uranium minerals from sea water or underground brines. Potassium occurs most commonly in clay minerals and feldspars, the latter being found in some sandstones as well as mudstones. Overall, mudstones are nearly always more radioactive than sandstones.

These radioactive elements emit gamma-rays as they decay. The rays are detected by a tool and recorded as a *gamma-ray log*. Specialised tools which can discriminate between gamma-rays emitted by uranium, potassium and thorium are used to record *spectral gamma-ray logs* which tell more about the mineral composition of the rocks than normal ones (diagram 9).

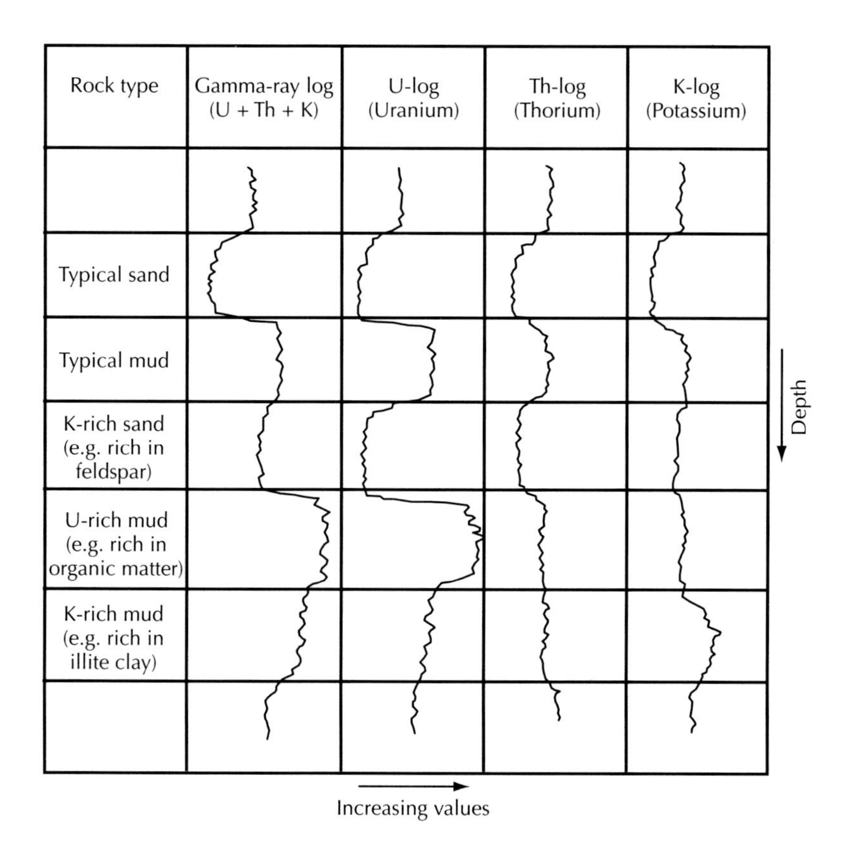

Diagram 9 Illustration of gamma-ray and spectral gamma-ray log response to different rock types.

The geologist also makes use of other special logs:

- *Neutron logs* record the amount of water in the rock, either free in rock pores or as bound water contained within clay minerals. Wet, porous sandstones and limestones as well as mudstones give high neutron readings. Porous rocks filled with gas and non-porous rocks such as evaporites give low neutron readings.
- The *density log* reads the bulk density of the rock (that is, the total density of the rock including that of the fluids contained in any pores). The densities of mudstones, porous sandstones and limestones are lower, those of non-porous sandstones and limestones higher. By comparing (overlaying) the neutron and density readings for each layer of rock, it is possible to work out what the logs can tell us about the composition of the rock (diagram 10).

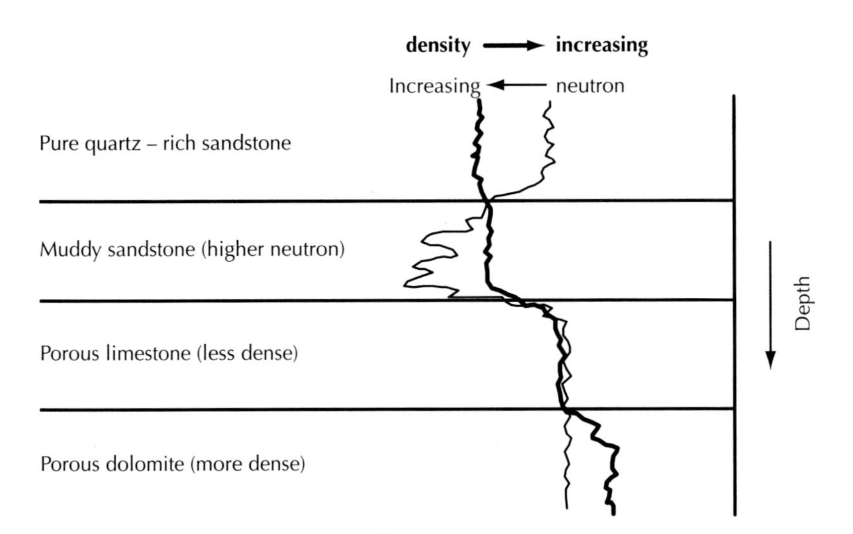

Diagram 10 Response of neutron and density logs to different rock types. The plots are scaled to coincide in limestone rocks. Dolomites (magnesium-rich carbonate rocks) are more dense than limestones. Pure sandstones are less dense, while muddy sandstones have similar density but higher neutron readings due to their mud content.

- *PEF (photoelectric factor) logs* record the atomic weight of the minerals in the rock. In general, mudstones and sandstones contain minerals with lower atomic weights than limestones. By combining the gamma-ray and PEF logs, one can distinguish these different rock types from each other (diagram 11).
- By means of the *sonic log*, one can also measure the time it takes for a shock-wave to pass through each layer of rock in the well. This relates to the compactness of the rock: hard, cemented sandstones, coals or limestones have shorter travel times than soft or porous mudstones and sandstones. The compactness of mudstones increases with burial depth so that a sudden increase in the overall compactness of mudstones can indicate a change to much older rocks which at some time in the past were uplifted from deep burial depths before being covered by much less compacted muds. The sonic log is commonly used to calibrate the results of a seismic survey, allowing the reflections identified by the survey to be matched with layers of rock in the well that have contrasting (hence, reflective) acoustic properties. One can also release seismic energy from inside the borehole and record its arrival at geophones deployed at the surface, or even in a nearby well, providing details of the structure of the layers around the well.

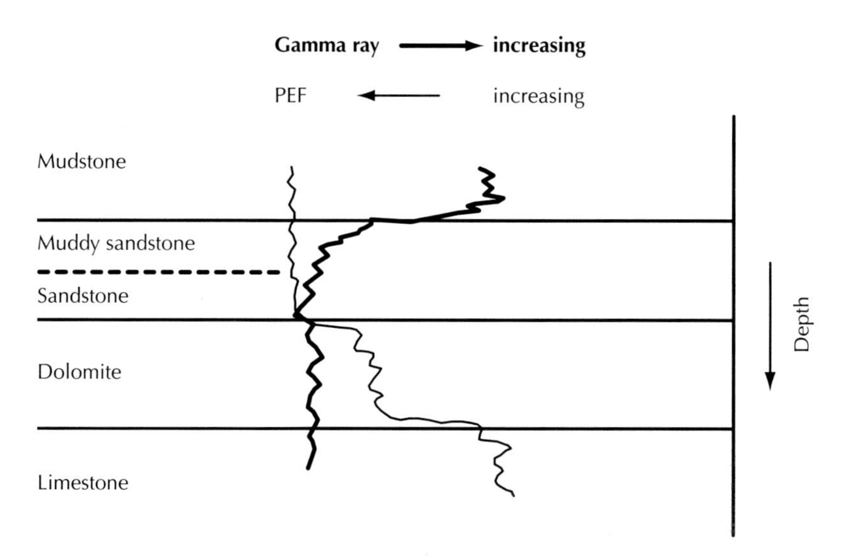

Diagram 11 Response of PEF and gamma-ray logs to different rock types. PEF is high in carbonate rocks, being higher in limestones than in dolomites. Gamma-radiation is high in mudstones.

A range of tools is available to measure the properties of fluids within the rocks in the well; in an oil or gas well; their main use is to assess the types and volumes of hydrocarbons present in the rocks rather than interpreting their depositional origin.

The logs recorded by these tools combine to help identify the general composition of the rock layers penetrated by a wellbore but cannot define sedimentary structures; cores are needed for that. Some details can be added by use of high-resolution imaging tools, which measure subtle variations in rock composition to produce an image of the borehole wall. This can reveal cm-scale cross-bedding, structural dip, large burrows filled with different sediment to the enclosing rock, pebbles and nodules of contrasting rock type. Unfortunately, such logs are expensive to acquire and are not generally obtained from most wells.

Interpreting the scale and impact of events

So far, we have looked at the ways sedimentary rocks are deposited, the properties of the rocks which we can measure to identify the processes by which they were deposited, and we have considered the main environments in which those processes operate. By studying a sequence of rocks, we can build up a picture of the environment in an area millions of years ago and begin to understand how that environment changed over time.

When considering such records of environmental changes, a number of questions come to mind:

- How important were these environmental changes to the development of life?
- How would similar events affect us now?
- To what extent are events, which we would regard as dramatic or disastrous if they occurred today, detectable in the geologic record?

The scale and frequency of events influences the likelihood of detecting them in sedimentary rocks. A historical example helps to put geologically significant events into perspective. During the late 1800s, the Indus river valley was affected by a major, catastrophic flood surging down from the glaciated Himalayan mountain range. The flood waters had been released from a lake in the high mountains which had been dammed by a mountain glacier; a minor earthquake broke the dam and drained the lake into the headwaters of the Indus River. A vast surge of water flooded down the narrow, upper reaches of the valley, wiping out a British Army punitive expedition which had been sent to quell a native uprising. The floodwaters spread out into the broad plains of the Indus Basin, depositing sediments which had been carried from the glacial lake and those eroded from the upper reaches of the valley by the rush of floodwaters (and probably also including a few Victorian military artefacts!). This event, recorded in the sediments of the upper Indus valley by a sequence of erosion and deposition, must have had a profound effect on the local people. Many events of a similar nature, occurring during the gradual deglaciation of the Himalayas since the end of the last Ice Age around 12,000 years ago, are doubtless recorded in the upper Indus valley sediments.

Such "glacial undamming" events are recorded very clearly by the drained lakes of the deglaciated mountain range. Over the course of thousands and millions of years, sediments accumulate in the lower reaches of the Indus valley as the valley-floor subsides. Since the total volume of rainwater falling all over the drainage area of the Indus river during monsoon floods far exceeds the volume of meltwater released by undamming of glacial lakes, the sedimentary record of the occasional undamming floods is totally overprinted; undamming events would have to be much larger in scale than our Victorian example to have a clear geological record in the sediments deposited in the Indus basin. A well known example of this is the creation of the "scablands" topography of the northern US during the sudden release of huge volumes of glacial meltwater from the Great Lakes 14,000 years ago, at the end of the last Ice Age (see Chapters 4 and 7). The effect of such an event on a populated area can scarcely be imagined. The channels and bars formed by these vast floods have been compared in scale to features seen on Mars at the Pathfinder landing site, although the Martian features are more likely to have formed by the melting of ice mixed in with sediments during some kind of heating event, rather than flow from pre-existing lakes.

This highlights two points. Catastrophic events which have a considerable local effect may go unnoticed on a regional scale. More importantly, the type and scale of events which can occur in any one place alter as regional conditions evolve – for example, as in the change from an unstable, deglaciating world to the more recent, relative stability of our present interglacial period.

Another important point to be borne in mind is that the geologic record is dominated by large-scale events representing only small periods of time. A couple of days of hurricane-force winds and waves will destroy the record of hundreds or even thousands of years of earlier calmer conditions. A landslide deposit 10 m thick, formed in a few minutes, may remain as an elevated lobe of sediment for thousands of years before being buried by younger sediments as a basin subsides. This may lead to an alarming impression of a series of disasters piled one upon another, even though the events responsible occurred at intervals of thousands of years.

In order to develop a realistic appreciation of the events which have shaped the Earth, as well as to predict future events, we need clear and careful descriptions of those events and an accurate idea of the age of the events and the time intervals separating them. This is the basic aim of *stratigraphy* (meaning "the description of layers"), the sub-division and description of sedimentary rock sequences in manageable sub-units.

In general, stratigraphy consists of the description of the rock record of past events as shown by the layers of sediment deposited over long periods of time. Distinctive events affecting a wide area are used to match up (correlate) the records from different places within this area. This procedure is similar to the practise of *dendrochronology* – the study of tree rings. Trees grow at different rates according to the local environmental conditions. Distinctive events, such as severe winters, are recorded by trees growing at the same time in widely separated places. The pattern of groups of severe winters and summer droughts can be recognised at a regional scale and, by analysing tree-ring patterns from very old and dead trees, the record has been extended back for many centuries. In rock stratigraphy, the distinctive events are recorded by changes in rock type at a much longer time-scale. Techniques for determining this time scale are described in the next part of this book.

4

Clocks of all sorts

The history of the Earth is, of couse, about what happened here in the past and when. Chapters 2 and 3 described how the measuring and recording of what happened are done, but we need also to understand when, and how quickly, significant changes in the Earth's environment did (and thus can) occur if we are to predict what may take place in the future. To recognise patterns of local and global change we need an accurate record of the order and speed of past events. Then we can try to deduce a progression of cause and effect between separate events.

Working out when things happened in Earth history is rather like dating events in our own lives. We need some kind of clock or calendar to measure the passage of time as well as a record of what the clock said the time was when particular events happened.

Types of clock

Earth history records two types of events – events which are regular and those which are irregular or "random". The regular events (days, months, years and lots of others) are used as the Earth's clocks: they mark off time and can be used to give absolute ages. They are regular because they result from the action of one or two simple physical laws rather than complex interactions, which give rise to irregular or chaotic systems. The two main types of clocks used in the study of Earth history are radiometric and astronomical.

Radiometric clocks

The rates of radioactive decay of certain unstable elements are absolutely regular when averaged out over a period of time; nothing is known to change them. To be useful as clocks, the elements have to be fairly common in natural minerals and to decay slowly over millions of years (they would be of little help if they decayed within, say, a few hours or days after formation) to form recognisable "daughter" products which are preserved in the same minerals. The main elements used are uranium, rubidium, potassium and carbon.

For example, an atom of radioactive rubidium decays to form an atom of strontium (another element) by converting a neutron in its nucleus to a proton and releasing an electron, generating energy in the process. We know from laboratory observations just how quickly that happens. The radiogenic daughter products of the decay – in this case strontium atoms – diffuse away and are lost above a certain very high temperature so, by measuring the exact proportions of rubidium and strontium atoms that are present in a mineral, we can work out how long it has been since the mineral cooled below that critical "blocking" temperature. The main problems with this dating method are the difficulty in finding minerals containing rubidium, the accuracy with which the proportions of rubidium and strontium are measured, and the fact that the method gives only the date when the mineral last cooled below the blocking temperature. Because the blocking temperature is very high, the method is used mainly for recrystallised (igneous or metamorphic) rocks, not for sediments – rubidium-bearing minerals in sediments simply record the age of cooling of the rocks which were eroded to form the sediments, not the age of deposition of the sediments themselves.

Potassium decays to form argon (a gas) which is sometimes lost from its host mineral by escaping through pores. Although potassium-argon dating is therefore rather unreliable, it can sometimes be useful in dating sedimentary rocks because potassium is common in some minerals which form in sediments at low temperatures. Assuming no argon has escaped, the potassium-argon date records the age of the sediments themselves.

Carbon dating is mainly used in archaeology. Most carbon atoms (carbon-12) are stable and do not change over time. However, cosmic radiation bombarding the upper atmosphere is constantly interacting with nitrogen in the atmosphere to create an unstable form of carbon, *carbon-14*. This radioactive form of carbon decays rapidly in geological time to become nitrogen again which, of course, is a gas. The atmosphere contains a mixture of carbon-12 and carbon-14 which remains fairly constant, the loss of carbon-14 via radioactive decay being balanced by the formation of new C-14 by cosmic rays. Living organisms, which exchange carbon with the atmosphere constantly through respiration and waste excretion, contain a similar mix of carbon-12 and carbon-14 to that of the atmosphere *at the time they were alive*. However, when they die and stop exchanging carbon with the atmosphere, the amount of unstable C-14 trapped within their bodies steadily decreases as it decays to form nitrogen. So, the proportion of C-14 to original C-12 in a dead organism decreases with time after death.

The *half-life* of carbon-14 is 5,730 years – that means that half the radioactive C-14 atoms present in a sample decay to form nitrogen after 5,730 years. During the next half-life, half the remaining radioactive atoms decay, half of *that* remainder in a further 5,730 years and so on and on. After ten half-lives (57,300 years) only one of every 1,024 of the original carbon-14 atoms will be left. The process is therefore too rapid to be of use in dating carbon compounds older than about 100,000 years because so little of the radioactive carbon will be left after such a long time that it is impossible to measure it accurately. Of course, 100,000 years is extremely important in the context of human history, so for human remains and artefacts carbon dating is extremely valuable. The half-life of other radioactive elements is much longer – more than 4,000 million years for rubidium and uranium, and over 100 million years for potassium – which makes those elements more useful as Earth clocks.

The accuracy of radiometric dates is the subject of considerable debate but most of the radiometric techniques claim an accuracy to within one or two per cent. This is fine for establishing the general age of the Earth, for example, at about 4,550 million years, and it can be used to give an accurate age for some events which are closely associated with the formation of igneous rocks. For example, the extinction of the dinosaurs is believed to have resulted from a collision between the Earth and a comet or an asteroid, and the time can be calculated to within 100,000 years by dating minerals formed in the impact crater. But the radioactive elements found in sedimentary rocks generally do not record the date of their deposition, so direct radiometric dating is usually impossible for the events in Earth history recorded by those rocks.

Similar in some respects to radiometric dating is a technique for dating the mineral *apatite* (calcium phosphate). Apatite commonly contains small inclusions of radioactive material. The radioactive decay gives off high-energy rays which damage the structure of the apatite crystal, causing linear "fission tracks". If the apatite crystal is heated above about 100°C, the tracks begin to heal up or *anneal*. When it cools again, new tracks start to form, so the density of tracks gives an indication of the length of time since the apatite last cooled below the annealing temperature. Apatite is quite common in igneous rocks; it occurs in sediments because of erosion of igneous rocks and because of the presence of fossil teeth and bone. The technique is restricted to dating apatite crystals eroded from igneous rocks because only these contain radioactive inclusions and show fission tracks. The annealing temperature is low enough for apatite grains in sedimentary rocks to anneal when the sediments are buried to a depth of a few thousand metres; new tracks form as the sedimentary rocks are uplifted towards the cooler surface. Apatite fission track analysis is therefore used mainly to record the history of burial and uplift of rocks in a particular area, and to get an idea of the temperatures during the time of burial.

Astronomical clocks

Astronomical clocks make use of regular changes in the orbital and rotational behaviour of the Earth, which result from its gravitational interaction with the Sun, the Moon and the other planets. They occur with a range of time scales. The most obvious, and shortest, of these effects is the day. There are also twice-daily changes in local sea-level due to the tidal pull on the oceans by the gravitational attraction of the Sun and Moon. More subtle variations follow from the 28-day orbital period of the Moon around the Earth, which affects the strength of tidal currents and influences some biological processes.

The amount of daylight experienced in parts of the world changes throughout the year, which is one of the factors causing the climatic variations we call *seasons*. The seasons are fundamental to much of the rest of this book and contribute to revolutionary new insights into Earth history. They are controlled and modulated by a number of astronomical processes which are listed below in order of increasing time-scale:

- annual repetition of the seasons
- decadal and millennial enhancement of the seasons (including 11-year sun-spot cycles)
- precession-induced extremes of seasons (20,000-year scale)
- obliquity-induced extremes of seasons (40,000-year scale)
- eccentricity-induced extremes of seasons (100,000-year scale).

We experience seasons because the Earth's axis of rotation is not parallel to the axis of its orbit around the Sun but is currently tilted by about 23.5°. In the northern hemisphere's summer, the North Pole is tilted towards the sun. The sun is more directly overhead and the temperature is generally higher. Six months later, the Earth has travelled round to the other side of the orbit so the North Pole is tilted away from the sun. Winter reigns in the north, the sun shines from lower down in the sky and it becomes colder. Just the reverse happens in the south, which is why Australians and New Zealanders can celebrate Christmas on the beach while most of the Europeans and Americans traditionally huddle round the fire (diagram 12).

If, instead of being tilted by 23.5°, the rotational axis had no tilt and was parallel with the orbital axis, all points on the Earth would experience 12 hours of daylight throughout the year and there would be no seasons (of course, it would still be hotter at the equator than at the poles but there would be no variation during the year as there is now). At the other extreme, if the rotational axis were tilted at 90° to the orbital axis, one pole would be in permanent daylight and the other in permanent night. This is virtually the case for the planet Uranus which has an axial tilt of 86° – although it is so far

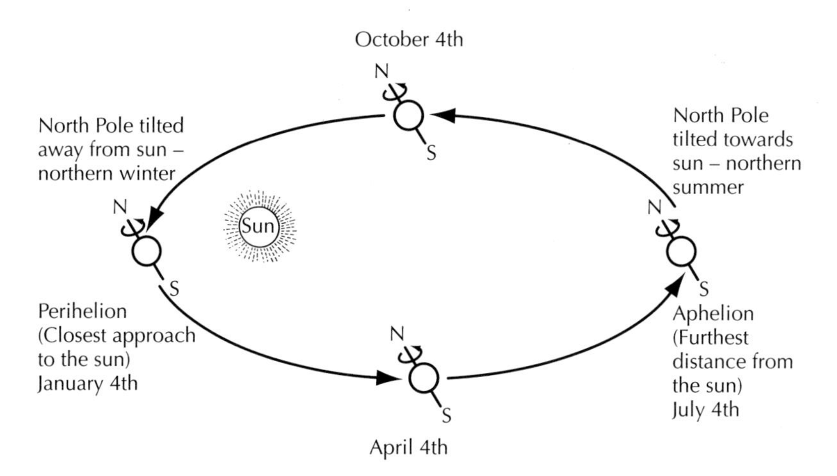

Diagram 12 The tilt of the Earth's axis is towards the sun in summer. The Earth's orbit is elliptical, bringing the Earth closest to the sun (perihelion) during the summer in the southern hemisphere.

from the sun that this probably doesn't make a lot of difference to surface temperatures. But on Earth, the seasons are extremely important. Perhaps most crucially, they control the melting of fallen snow around the polar regions, so governing the tendency for ice caps to accumulate and controlling the reflectivity of the surface. Snow is much more reflective than snow-free land or water.

The reflectivity has an effect on the amount of sunlight trapped as heat energy by the Earth's surface so that, as snow cover builds up, less heat is trapped and the Earth's atmosphere cools. On the other hand, when permafrost around the edges of the ice-caps melts, it releases the hydrocarbon gas methane from decaying vegetation into the atmosphere. Methane is one of the so-called "greenhouse gases" which contribute to global warming. Other similar positive feedbacks which enhance global warming involve the release of more carbon dioxide and methane by biological processes, and the release of water vapour into the atmosphere to form clouds, which also have a "greenhouse" blanketting effect.

The amount of seasonal climate variation depends on two factors: the tilt of the rotational axis (currently about 23.5°, known as the *obliquity of the ecliptic* or simply *obliquity* for short) and the *eccentricity* (ovalness) of the Earth's orbit. Because of the competing pull on the Earth of the Sun, Moon and planets, the obliquity (tilt) of the Earth's rotation axis gradually varies between about 22° and 24°, a full cycle of this "nodding" motion taking about 41,000 years to complete.

The shape of the Earth's orbit varies from near-circular to more elliptical, with extremes being repeated about every 100,000 and 413,000 years. If the Earth's climate were not seasonal, this would make little difference. But with a tilted rotational axis, summer warmth and winter cold are affected by how close the Earth is to the Sun during summer and winter. With a near-circular orbit, there is little contrast in seasons due to equal distance from the Sun at all times but when the orbit is highly elliptical the position of the Earth on its orbit in summer and winter is highly important, a phenomenon known as *precession* (diagram 13).

Precession (or more exactly *the precession of the equinoxes*) is caused by a rolling wobble of the rotation axis which goes around in about 26,000 years; this wobble is similar to the roll of the axis of a spinning top or gyroscope. The effect is to change the direction in which the tilted axis points. At present, the North pole points directly towards Polaris (the Pole Star) in the constellation Ursa Minor. For some of the precession cycle, the North Pole will be tilted towards the Sun (i.e. northern hemisphere summer) when the Earth is near *perihelion*, the point in its orbit closest to the Sun. The northern hemisphere summer is unusually warm during this period – or to be more accurate, there is more *insolation*: more of the Sun's radiation reaches the top of the atmosphere during that summer. But in later years, due to precession, the axis will be pointing directly at the Sun (i.e. at northern midsummer) when the Earth has reached a point slightly further around the orbit, slightly further away from the Sun. At present, the Earth is closest to the Sun in the northern hemisphere's winter (early January). Given that the precession effect takes 26,000 years to cause the Earth's axis to move round in a circle, one would expect that once every 26,000 years (orbits), the axis would be pointing directly at the Sun at perihelion. However, due to the changes in eccentricity mentioned above, the shape of the orbit is also changing at the same time. This causes the position of perihelion to change over thousands of years. The overall result is that the Earth's axis points directly at the Sun at perihelion earlier than expected. What this means is that on average every 21,000 years there is a change from warmer summers to cooler summers and back again. These changes due to the precession effect are greatest when the orbit is most elliptical, so the precession effect is enhanced by the 100,000-year and 413,000-year eccentricity cycles (diagram 14).

These changes in obliquity, eccentricity and precession are called *Milankovitch cycles* after a Serbian scientist who studied them in the early 1940s. He calculated the insolation at various points in the cycles and predicted how this might have affected climate during the last Ice Age. This gave added credibility to the ideas of some Victorian geologists that astronomical cycles might be recorded by changing patterns of deposition in sedimentary

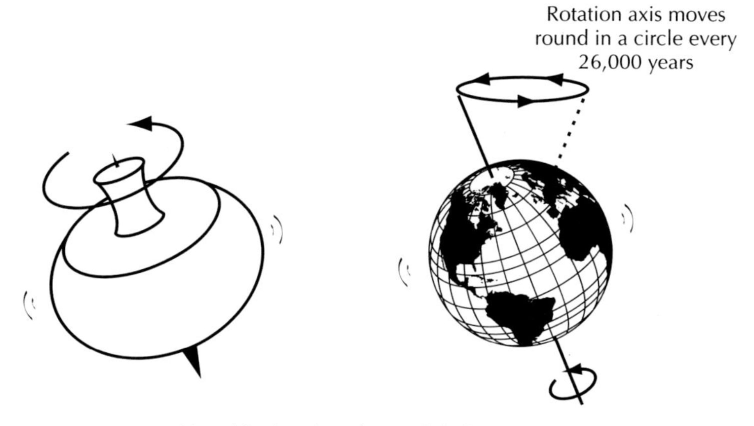

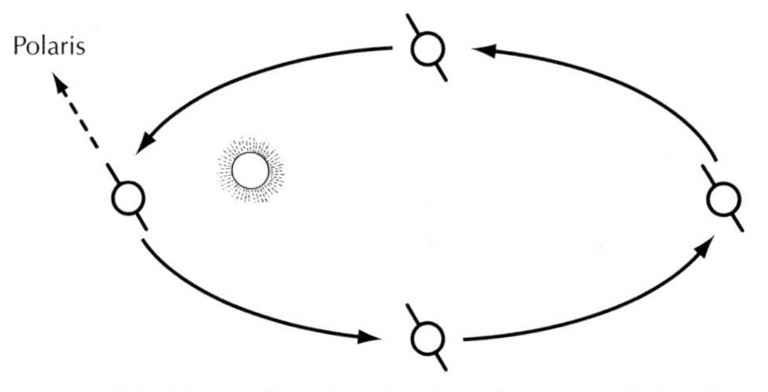

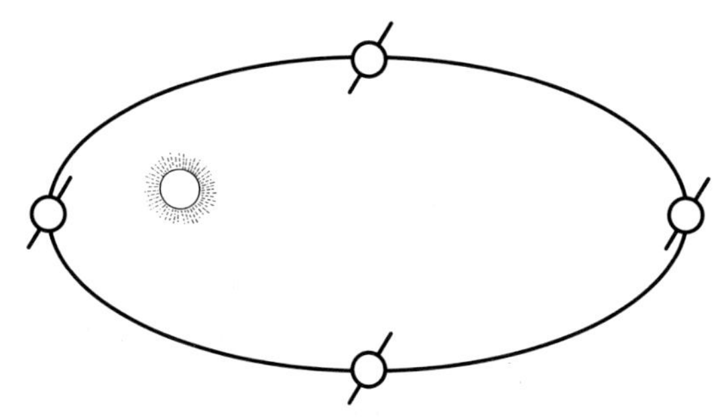

Diagram 13 The Earth's axis wobbles in a circle every 26,000 years. This causes the timing of the seasons to change, relative to the position of the Earth in its orbit.

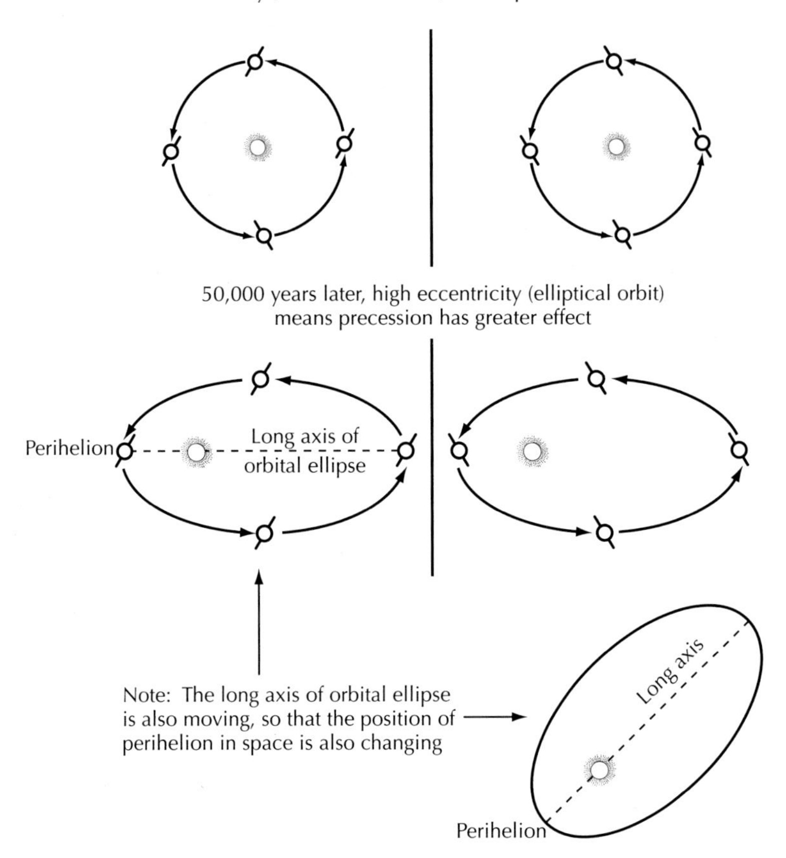

Diagram 14 The shape of the Earth's orbit varies from more circular to more elliptical. This determines how much the precession effect influences the Earth's climate.

rocks. Although it has so far been difficult to calculate with any certainty the actual amount of insolation before about 5 million years ago, astrophysicists generally agree that the Milankovitch cycles have changed little over the past few hundred million years. The eccentricity cycles, in particular, have been constant for at least 400 million years, while the precession and obliquity cycles have been gradually slowing down through time as a result of tidal friction between the Earth, Moon and Sun.

It is clear that throughout the last few hundred million years the amount of insolation has varied regularly, causing regular extremes of climate and seasonal contrast. We need to know what effect has this had on the Earth's surface and how it has been recorded in Earth history. This information is vital because, if we can detect regular events marking off periods of 20,000 to 400,000 years, we will have a dating tool of greater accuracy than radiometric clocks, one which might be of much wider value in dating sedimentary rocks.

By far the best known example of how Milankovitch cycles have affected the Earth's climate and environment is their influence during the last Ice Age. There has been a major ice cap at the South Pole for at least 30 million years but severe glaciation of both northern and southern hemispheres began only at the start of the Pleistocene Epoch, about 1.8 million years ago. Northern and southern polar ice sheets several kilometres thick built up rapidly, advancing and retreating periodically as the Earth underwent colder "glacial" and warmer "interglacial" episodes. In ice-free latitudes, these episodes are recorded as drier and wetter (or pluvial) phases. Periods of cold, dry climate alternating with warm, wet climate are recorded by the *loess* (wind-blown dust) deposits of China.

The ice in the polar ice caps consists of water which evaporated to form clouds above the oceans, travelled to the poles and fell as snow. These physical transport processes affect the composition of the water that eventually falls as snow. This is because all water contains oxygen, an element which occurs in three different forms, or *isotopes*: all have the same chemical characteristics but each contains a different number of neutrons in the nucleus of its atom. *Oxygen-18*, the isotope with most neutrons, is over ten per cent heavier than the one with fewest neutrons, oxygen-16. These isotopes are all stable and do not disintegrate by radioactive decay.

Lighter water molecules are more readily evaporated from the ocean's surface. Thus, water in clouds contains a higher proportion of oxygen-16 than that in the ocean. On its way to the pole, the cloud loses some of its water as rain in higher latitudes, so that oxygen-18 water molecules are preferentially lost as they are more readily precipitated as rain on their way to the poles. Hence, the water that eventually falls at the poles as snow contains a high proportion of the lighter oxygen-16. When polar ice melts, the meltwater contains much more oxygen-16 than ordinary seawater. Because it is cold and therefore dense, polar meltwater tends to flow along the bottoms of the oceans. Microscopic creatures living at the bottom of the sea incorporate oxygen from this polar meltwater into their shells. Drill cores from ancient deep-sea sediments in the Atlantic and Pacific Oceans show almost identical changes in the oxygen isotope composition of these shells through the last 2.5 million years, changes responding to variations in the amount of water melting from the ice caps (diagram 15).

This record has now been supported by direct evidence of ice-cap composition obtained from drill cores of polar ice. Provided the age of the cored

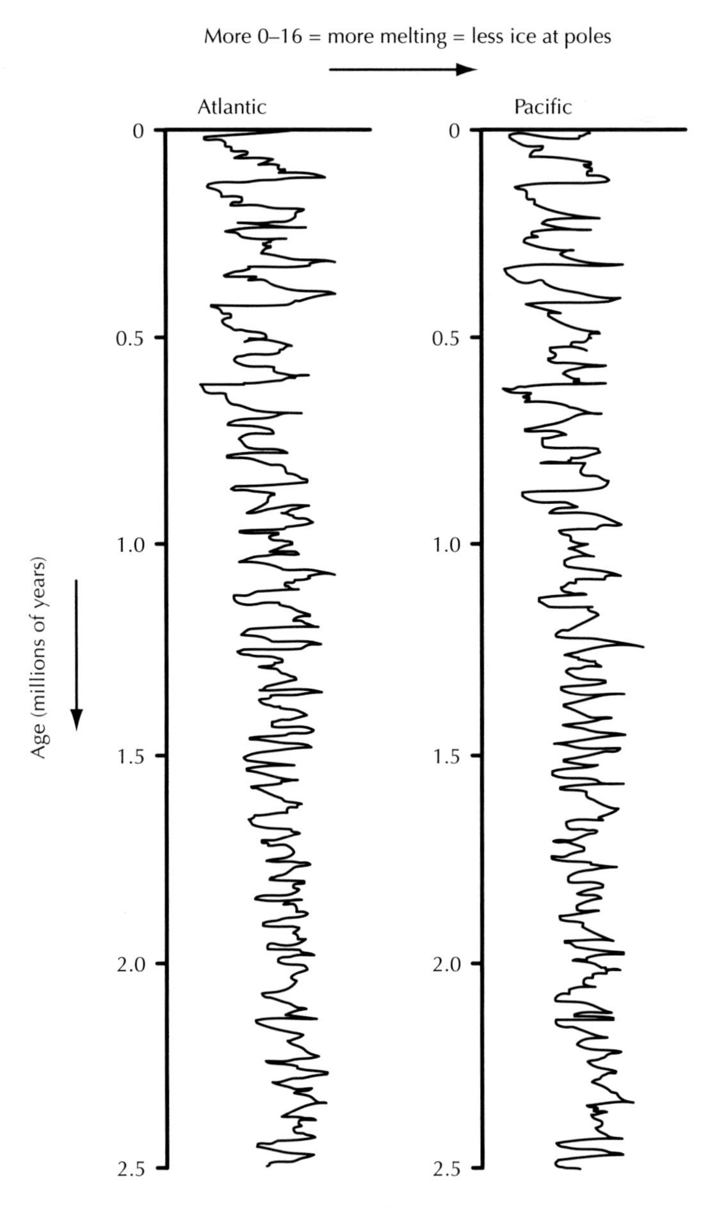

Diagram 15 Variations in the proportions of Oxygen-16 and Oxygen-18 in the shells of ocean-floor organisms in the Atlantic and Pacific oceans during the last 2.5 million years. The compositions of the shells varied in step, due to periodic O-16 input into the world's oceans from melting polar ice caps.

samples is known, both sets of records can be analysed for periodic changes over time. The age control is provided by radiometric dating of the sea-floor basalts underlying the sediments and by the record of magnetic reversals recorded by the sediments themselves (the cause and dating of magnetic reversals is discussed in detail later in this chapter). The results of the analyses show that, throughout the last 2.5 million years, the sea-floor records from the Atlantic and Pacific Oceans were controlled by the 41,000-year obliquity cycle and the 100,000-year eccentricity cycle, although the 100,000-year cycle had less influence on polar ice volumes before 900,000 years ago (diagram 16).

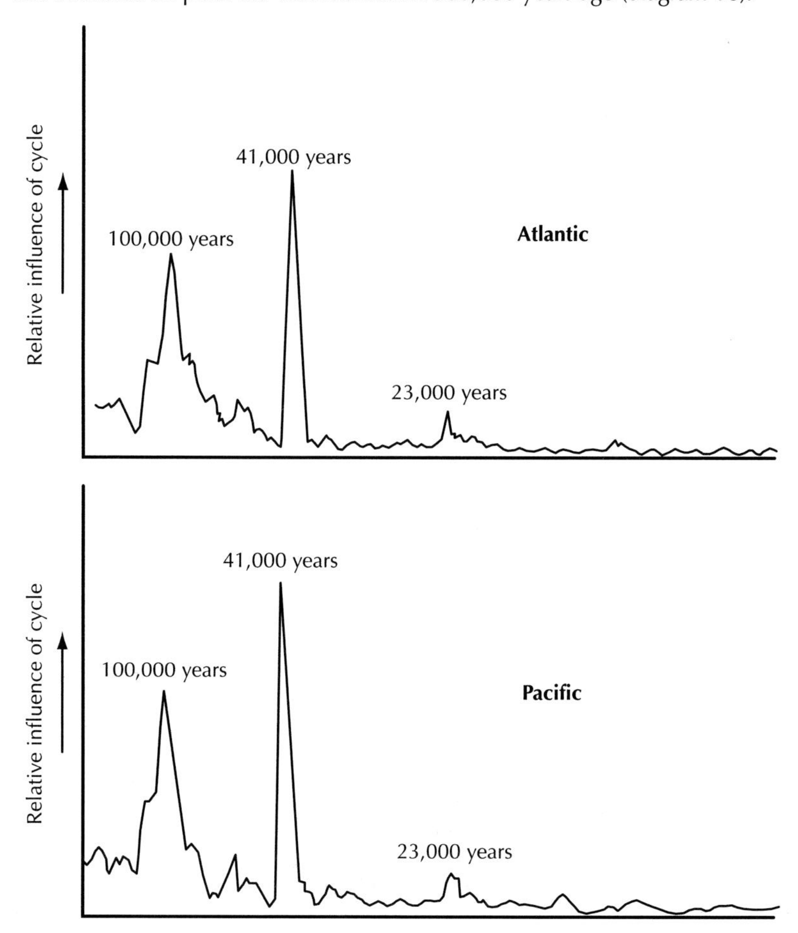

Diagram 16 Analysis of the O-16 records in diagram 15 shows that the data from the two oceans is strongly cyclic, with the dominant variation being due to changes every 100,000 years, 41,000 years and 23,000 years.

This fact has convinced most Earth scientists that the pattern of climate changes predicted by Milankovitch strongly affected the Ice Age Earth over the last few million years. Recent research has shown similar patterns of change recorded in the Chinese loess deposits during the same period. More excitingly, work on ocean floor sediments in the Equatorial Atlantic demonstrates that Milankovitch-scale patterns of climate change have been going on for at least the last 30 million years. It is less clear exactly how these astronomical cycles have caused the changes in deep-sea sediment composition, and relatively few geologists expect such patterns to be clearly recorded before the growth of polar ice caps about 30 million years ago.

There have been some intriguing attempts to demonstrate a record of Milankovitch climate cycles in older sediments but they are generally unconvincing because it is so difficult to establish the exact duration of periodic events which happened so long ago. Without any other way of dating events to an accuracy of better than 100,000 years, how can we know that a periodic change in the environment occurred every 100,000 years (and thus may represent an eccentricity cycle) or every 140,000 years (and so had some other origin)? Without such proof, the idea that Milankovitch climate cycles can be used to provide a high-resolution time-scale in rocks older than a few million years is an elusive dream. We will return to this dream at the end of Chapter 5 – and see its realisation in the latest oil-industry studies of Earth history in Chapter 6. But first we have to consider other ways of marking out time in Earth history, by the study of *stratigraphy*.

5

Events as time markers

The absolute age of an event in Earth history can be measured with one or more of the Earth's clocks. In most cases this is very difficult, either because the clocks are inaccurate (often the case in radiometric dating) or are difficult to read (almost always true for Milankovitch climate cycles); indeed, in many cases we can't be certain that the event in question occurred at the same time as the clock we are using was set.

As an alternative, we can look for distinctive, widespread events of any origin and use them to build up a sequential pattern of markers. The absolute age of such indicators may not be measured directly but we can often work out their time sequence. The "age" of each marker in the sequence is then given not in years but by reference to a letter code or name (as in "the Jurassic Period"). The relative age of any other event can then be determined by reference to its position in the sequence (as in the "late Jurassic"). The ultimate goal is to define a sequence of regional or global events and eventually tie it into an absolute time framework by careful correlation with the Earth's clocks, something which is now beginning to be done for the whole of Earth history. At present, the major events are dated by reference to radiometric ages. For example, we think that the end of the Jurassic Period occurred about 142 ± 2.6 million years ago, from radiometric dating of volcanic rocks which were erupted at about that time (diagram 17).

Biostratigraphy

The most useful markers are ones which occur commonly and which can be identified easily. Fossils are a good example. They record the evolutionary sequence of life forms over long periods. Careful studies since Victorian times have documented the sequential replacement of one fossil species by another as evolution proceeded, so providing a relative time scale with which we can date the rocks in which the fossils are found. The most useful fossil species for dating purposes are ones which had a global distribution for a short period and then evolved into distinctly different species. This science goes by the name of *biostratigraphy*.

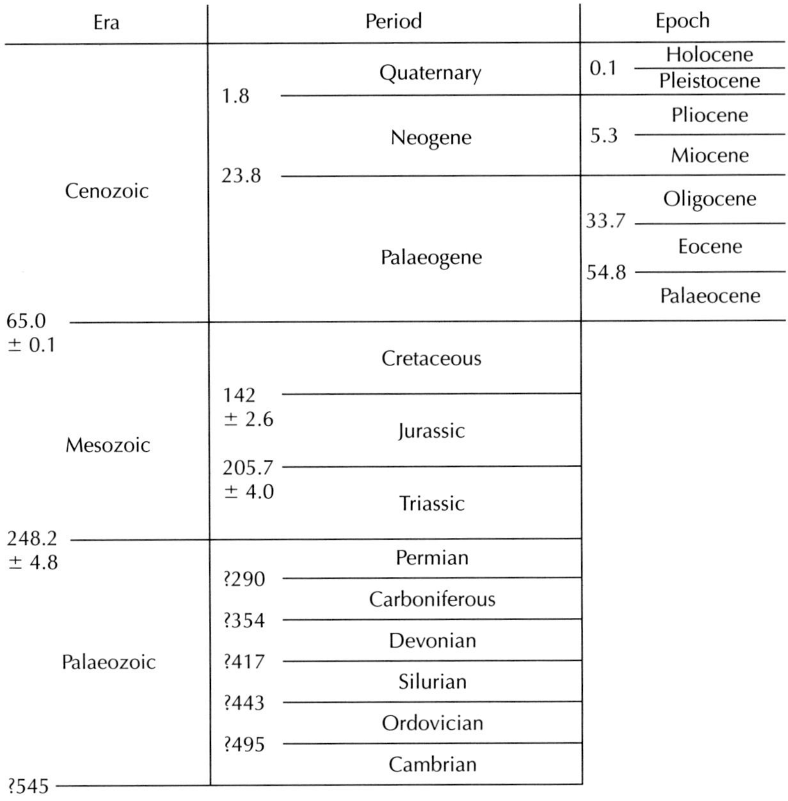

Era		Period		Epoch
Cenozoic		Quaternary		Holocene
			0.1	
				Pleistocene
	1.8			
		Neogene		Pliocene
			5.3	
				Miocene
	23.8			
		Palaeogene		Oligocene
			33.7	
				Eocene
			54.8	
				Palaeocene
65.0 ± 0.1				
Mesozoic		Cretaceous		
	142 ± 2.6			
		Jurassic		
	205.7 ± 4.0			
		Triassic		
248.2 ± 4.8				
Palaeozoic		Permian		
	?290			
		Carboniferous		
	?354			
		Devonian		
	?417			
		Silurian		
	?443			
		Ordovician		
	?495			
		Cambrian		
?545				

Note: Dates after the Mesozoic are probably correct. Mesozoic dates have a known uncertainty range.

Diagram 17 The geological time scale showing the ages of geological periods, determined from radiometric dating.

Ammonites are an excellent and well-known example of biostratigraphically useful fossils. They lived in the oceans of the Mesozoic Era between about 250 million and 65 million years ago when the dinosaurs walked the Earth. After death, their shells commonly floated for a while, so their remains became widely distributed in marine sediments. Ammonite species evolved rapidly into distinctly different forms, individual species lasting for about 500,000 years, and they have become the main index fossils for much of the Mesozoic period. Unfortunately, they are not really all that common. They tend to be abundant in thin beds which were deposited slowly, separated by thick sequences of more rapidly deposited sediments, so they can be difficult to find. I have carried out field surveys of Mesozoic rocks across the world for over ten years and, apart from a few localities where they are present in profusion (such as Lyme Regis in Dorset, visited by every English geology student, and a remote valley in the Himalayas where the local people were mining them as curios – believing them to be the petrified horns of ancient goats), in all that time I have found only a dozen or so. I probably wasn't looking in the right places. In any case, they are unlikely to turn up in drill cores which are typically 10–15 cm in diameter, so they are of relatively little use to oil companies and in deep-sea drilling research.

Microscopic fossils are generally much more abundant than larger ones like ammonites so they are often found in drill cores and borehole cuttings, and are accordingly more useful to oil companies and others as dating tools. Some types of microfossil evolved rapidly into readily-distinguishable species with a global distribution; good examples include the planktonic foraminifera, which have drifted at the surface of the oceans throughout the world since early Cretaceous times, over 130 million years ago. But these, too, have their problems:

- such tiny fossils commonly fall into boreholes from higher levels due to "caving"
- some species survived for very long periods, and
- some forms evolved which resemble others that existed earlier in the Earth's history.

Still, the abundance of these microfossils in marine rocks means that it is fairly easy to date marine sediments from an assemblage of different species of microfossil, and hundreds of individuals may be present in one fist-sized sample. The species usually have overlapping age ranges, giving a fairly precise date to the sediment in which they are found.

In non-marine sediments, it is generally much more difficult to find biostratigraphically useful fossils because most organisms adapted to life on land are hardy specialists, suited to a relatively tougher life-style than that of their oceanic cousins. Land species tend to survive for longer without evolving into distinctive new forms, something particularly true of microscopic fossils like the spores and pollens of plants found in non-marine sediments. However, spores and pollen give a good idea of the kind of vegetation surrounding the site of deposition and their records in well and outcrop sections are extremely useful in studying ancient climate changes.

Magnetic stratigraphy

The next most commonly used form of relative age dating is *magnetic stratigraphy,* the study of the history of reversals of the Earth's magnetic field. The Earth has a magnetic field due to the constant "boiling" motion of the molten iron in its outer core, coupled with its rotation. The axis of the magnetic field is currently aligned so that the *magnetic* North pole is located in northern Canada, hundreds of kilometres from the *rotational* (geographical) North pole. Paradoxically, what we call the "North" magnetic pole is by one definition a magnetic south pole! This is easy to understand – opposite poles of magnets attract each other, so the North pole of a compass needle actually points "North" because it is attracted to the Earth's "South" magnetic pole . . . or maybe we should rename the poles of the compass needle!

Whatever one calls the Earth's magnetic poles, they do not stay put over long periods of time. Due to variations in the patterns of upwelling in the molten outer core, the polarity of the magnetic field can sometimes reverse, so that the North magnetic pole flips round to the southern hemisphere where, via a series of erratic swings, it eventually settles down close to the South rotational pole.

The effects of such a magnetic reversal on modern technology would be dramatic. Apart from the disastrous effects on compass navigation, there would be a long period when the magnetic field collapsed, allowing high-energy radiation from the Sun and stars to penetrate the atmosphere. This would seriously affect radio and TV communications, and might affect computer chips in the same way as the electromagnetic pulse associated with atomic explosions. The high levels of radiation would also affect life on Earth in unpredictable but undoubtedly harmful ways. Fortunately, magnetic reversals occur on average only once in about 500,000 years, but their causes are not well understood and there are some researchers who think the next reversal may occur in the near future.

Whatever the future may hold in store, the study of magnetic stratigraphy allows one to record the pattern of past magnetic reversals. Newly-crystallised iron minerals record the prevailing polarity of the Earth's magnetic field (i.e., which end of the Earth is "North") at the time of their formation. Some iron minerals preserve this record better than others – the best are in volcanic lava flows like the ones formed at the mid-oceanic ridges. These are areas of almost continuous lava eruption at long lines of weakness in the Earth's surface where oceanic crustal plates are spreading apart to form new crust. The cooling lava records the magnetic polarity at the time of eruption and, as the plates spread apart, a conveyor-belt of cooled lava is formed on either side of the ridge. Polarity reversals result in bands of normal- and reverse-polarised lava, the lavas closest to the spreading axis being the youngest. The magnetism of these rocks is strong enough to be detected by ships passing above the ridges so it is relatively easy to document the record of polarity reversals as one travels away from the ridges and hence back through time. The absolute age of the magnetised lava can also be measured by radiometric dating of drill cores from the ocean floor. This has provided a record of magnetic reversals calibrated to an absolute time scale for the last 200 million years, which is the age of the oldest known ocean floor lavas.

Magnetised iron minerals also occur in sedimentary rocks, either as transported grains of iron mineral (which may align themselves parallel to the prevailing magnetic poles) or, more usefully, as crystals of iron oxides which formed in the sediments soon after deposition and inherited the existing magnetic polarity. The polarity of such minerals can be detected by taking orientated samples from rock outcrops or drill cores and measuring their magnetic polarity in the laboratory. A large number of samples from a sequence of rocks of progressively younger age allows one to compare the magnetic reversal pattern recorded by the rocks with that established for the ocean floor lavas.

For example, in a study I organised in southern Nepal, 128 samples of sedimentary mudstone were collected from a road cut through about 4,000 m thickness of non-marine sediments. The base of the section, where the oldest rocks were exposed, was known from fossil elephant, giraffe and cow bones to be about 10 million years old. The samples showed a long section of "normal" magnetic polarity near the base of the sequence, followed by a series of shorter reversals (diagram 18). The long-normal was matched to a known long-normal episode which lasted from 10.5 million to 9.5 million years ago, so we could correlate the later reversals with the known pattern of subsequent reversals taking place in the following 8 million years. The precise age data provided by this magnetic stratigraphy were later used in an investigation of important changes in climate and vegetation in Nepal, Pakistan and East Africa during the early evolution of Man, which are described further in Chapter 6.

Stable isotope stratigraphy

Earlier we looked at ways of dating events during the last Ice Age by measuring the proportions of stable oxygen isotopes in fossil shells and ice. Oxygen is not the only commonly-occurring element with a number of stable isotopes.

In addition to the unstable, radioactive form carbon-14 mentioned before, carbon has two stable forms, carbon-12 and carbon-13. Carbon-13 has one more neutron in its nucleus than carbon-12, and is therefore slightly

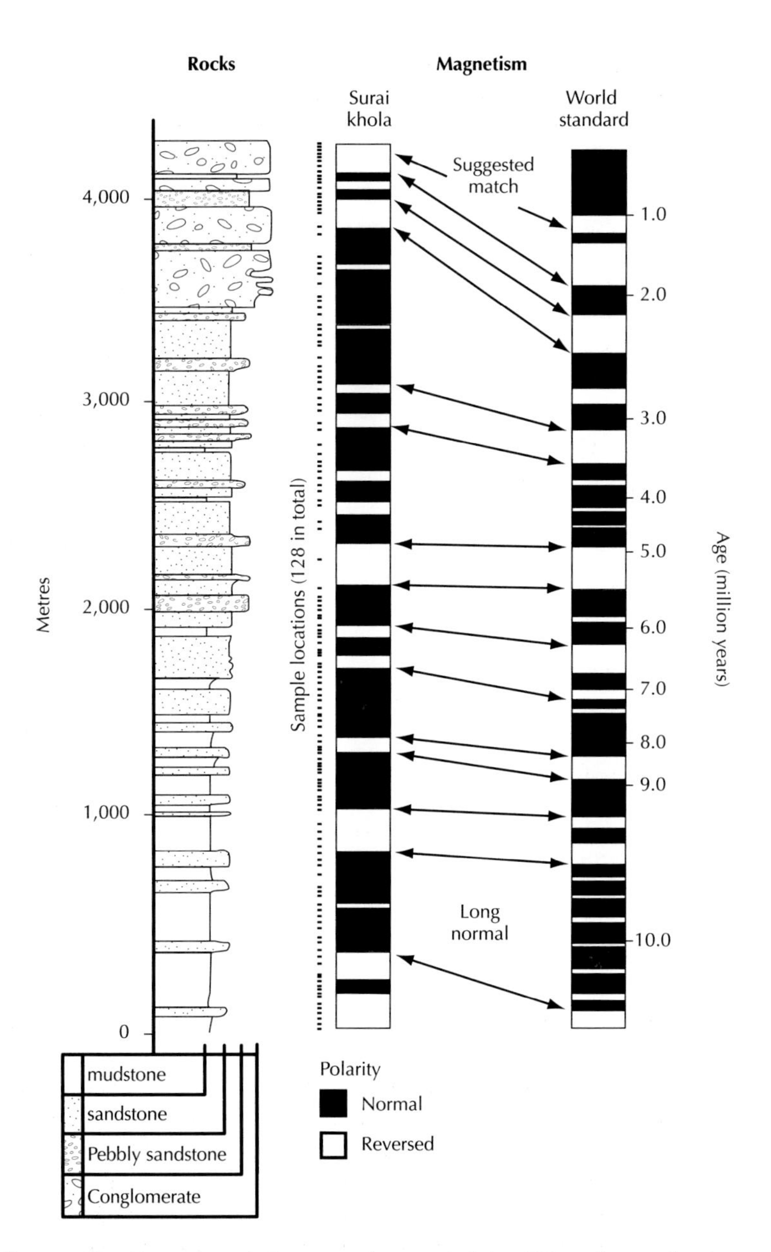

Diagram 18 Magnetic polarity reversals recorded by 128 rock samples from a sequence of sediments in Nepal. The approximate age of the sediments was already known, allowing the reversal record to be matched precisely with known past reversal events. This gives accurate dates for several layers in the rock succession.

heavier but has exactly the same chemical characteristics. Just as with the stable oxygen isotopes, this means that processes involving physical transfer of carbon result in differential sorting of the two stable carbon isotopes. Different processes result in a different degree of sorting, leaving a fingerprint of their operation on the carbon-bearing compounds that form as a result of the processes. This is particularly true of biological processes.

Plants take more of the lighter carbon-12 than the heavier carbon-13 from the atmosphere and store it in their tissues. Bacteria concentrate carbon-12 even more. By measuring the carbon isotope composition of calcium carbonate (limestone) in rocks, we can determine whether the limestone formation resulted from plant growth (as in soils) or bacterial processes (as in limestone cements created during shallow burial of sediments, when bacteria feed on buried organic matter in the sediments). An example of the use of carbon isotopes to trace plant evolution in the past is given in the next chapter.

Marine organisms trap carbon-12 in their bodies. When they die and are buried in marine sediments this carbon-12 is removed from the atmosphere-ocean system, until the sediments are uplifted to the surface and eroded. This cycle of burial and erosion of organic carbon is not always in balance. At some times in the past, huge quantities of organic carbon were rapidly buried in marine sediments, removing significant amounts of carbon-12 from the system. Heavier carbon-13 was therefore concentrated in the carbon remaining in the atmosphere and oceans. Marine organisms living in the carbon-13 rich seas formed shells rich in this isotope, so providing a record of those times. Given enough data, the patterns of the changing carbon-12 and carbon-13 content in the oceans as recorded by marine shells can be used as a dating method.

A similar approach to dating uses the relative abundance of strontium-86 and strontium-87 isotopes in marine shells. As we saw in our consideration of radiometric clocks, strontium-87 is formed by the radioactive decay of rubidium-87. It is formed in continental crust as igneous rocks cool and is then supplied to the ocean by erosion of these rocks at the surface. The proportions of strontium-86 and strontium-87 in the ocean waters at any time reflects the amount of erosion going on at that time, and has varied significantly during Earth history. Oceanic currents mix the water in the oceans quite efficiently so that at any point in time the strontium isotope composition of the ocean is much the same everywhere. Strontium-86 and strontium-87 are incorporated into the shells of living organisms (which are then buried in sediments) in the proportions present in the water in which the organisms live. Measuring the strontium isotope content of fossil sea-shells can therefore be a useful indication of their age, although chemical changes in the shells may occur after burial, altering the original isotope composition.

Sequence stratigraphy

Over the past few decades, seismic surveys (see Chapter 3) have been carried out on most of the world's coastal areas. They show the patterns of sediment layers that have been built up in coastal land areas and in the adjacent shallow seas. It was soon realised that the layers are not always flat and parallel but commonly show the advance of wedge-shaped piles of sediment out into the sea. This is evidence of sediment progradation, as explained in Chapter 3. Repeated episodes of progradation of sediments can be recognised on seismic records of the continental shelves, showing that huge amounts of sediment have advanced out from the land over the edges of the shelves, followed by episodes of drowning by the rising sea when the abandoned sediment wedges were covered by thin layers with more widespread extent (diagram 19).

These seismic patterns are clearly seen at the margins of the Gulf of Mexico, an oil-rich basin which was studied extensively in the 1960s and 1970s by the US oil company Exxon. Company geologists came up with an explanation of the seismic patterns in terms of the competing influences of sediment supply and changing accommodation space (as defined in Chapter 3). Episodes of sediment progradation were interpreted as times of increased sediment supply or lower accommodation space. The seismic records showed quite clearly that sediments deposited on the continental shelf, in

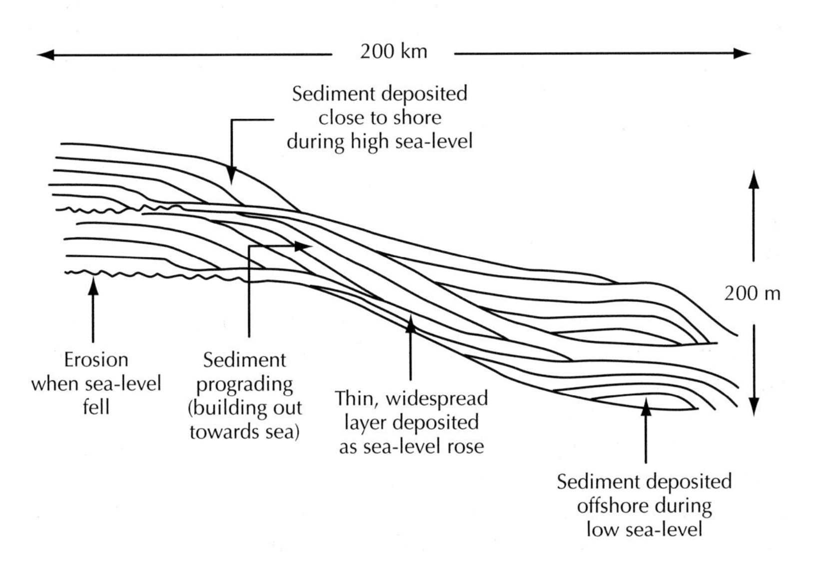

Diagram 19 Sketch showing how layers of sediment pile up on the continental shelf when sea-level is high, and are transported into deeper water when sea-level falls.

water depths of many tens of metres, had been eroded at times when sediment wedges built out into deeper water. The implication was that at these times the whole shelf must have been exposed to erosion. Fossils recovered from wells drilled through such sediments demonstrated that the most recent erosion events, which occurred during the last 2 million years, corresponded to times of major growth of ice sheets during the last Ice Age. It seemed obvious that the shelf had not been uplifted and subsided to this extent during the short time intervals involved. Instead, the growth of polar ice caps must have locked so much water up as ice that world sea-level fell, exposing the continental shelves to erosion and leading to the transport of eroded sediments into deeper water.

A huge database of drill cores and microfossil samples has now been assembled from this area. This shows beyond doubt that Ice Age sea-level fluctuations were responsible for the episodes of sediment progradation recorded by the seismic surveys. Generally, sediments were eroded and transported out into deep water at times of growing ice caps. The shelves were later flooded and draped with thin layers of sediments as the ice caps melted. Close to the Mississippi River delta, sediment progradation was out of step with world sea-level because sediment supply down the river increased as the North American ice caps melted and released glacial debris. Progradation close to the delta took place as other areas were flooded by rising sea-level. Within the valley of the Mississippi onshore, buried erosion surfaces record the incision of the river deep into the underlying bedrock at times of lower sea-level, followed by deposition of coarse river gravels as the ice caps melted.

The record of sediment progradation during the cold phases of the Ice Age has since been documented from seismic studies of the other continental margins. Major deltas such as the Amazon and Niger Deltas record times of minor deep-marine sediment input, when ice-caps retreated, followed by collapse and sliding of huge sections of the shelf into deep water as sea-level fell with the growth of polar ice sheets. These episodes have been dated by use of microfossils, the dates agreeing well with those from the Gulf of Mexico. Because the equatorial areas drained by the Amazon and Niger were not glaciated during the cold phases of the Ice Age, sediment supply down those rivers, unlike the Mississippi, did not increase during the melting of the ice caps. In fact, spore and pollen records from the Niger Delta suggest that the climate in that area was drier and perhaps cooler during the polar glaciations and wetter during melting of the ice caps, so to some extent sediment supply down the rivers probably did increase between glacial episodes, but much less so than in the Mississippi.

Surprisingly, similar patterns of sediment layers were seen on seismic records of pre-Ice Age sediments in the Gulf of Mexico. The southern US margin has built out into the Gulf gradually over hundreds of millions of years and the layer patterns show that this has occurred episodically. Just as in the Ice Age sediments, times of deposition on the flat shelf were punctuated by periods of widespread erosion of the shelf and building of sediment wedges into deeper water. Careful examination of the seismic records allowed the relative changes in water depth on the shelf to be estimated by measuring the point of *coastal onlap* (diagram 20). Fossils taken from wells drilled through the sediments gave an approximate date for each episode of sediment progradation and later flooding. Data from other parts of the world showed evidence of similar episodes of progradation, although the accuracy of dating is insufficient to prove beyond doubt that the events occurred everywhere at the same time. Despite this reservation, it was proposed that

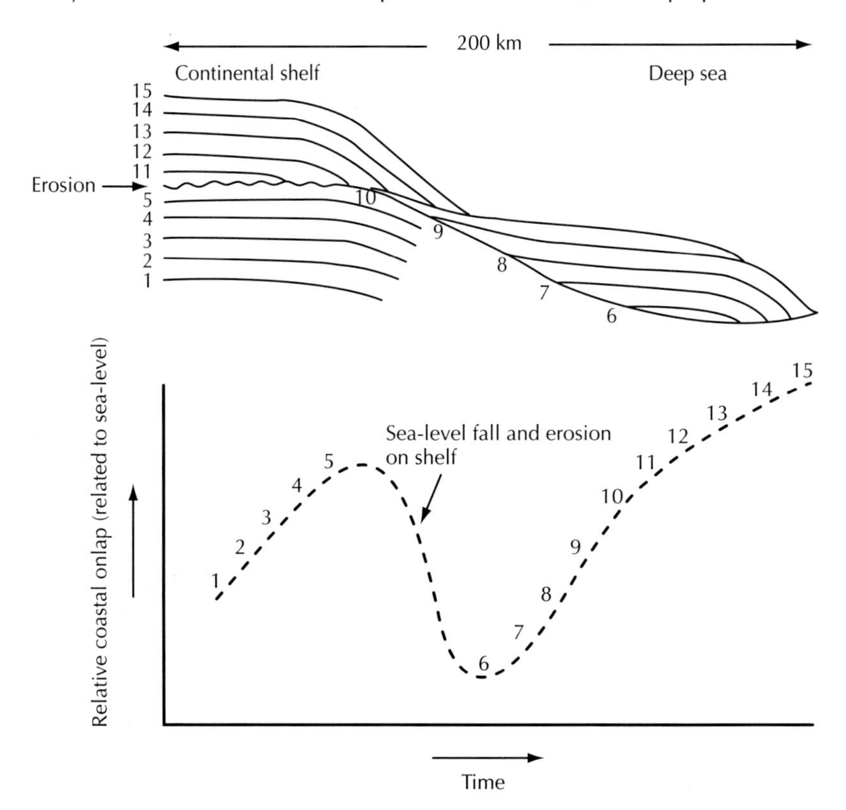

Diagram 20 Sketch showing how seismic records of sediment layering at the continental margins can be used to determine progressive changes in coastal onlap during cycles of rising and falling sea-level.

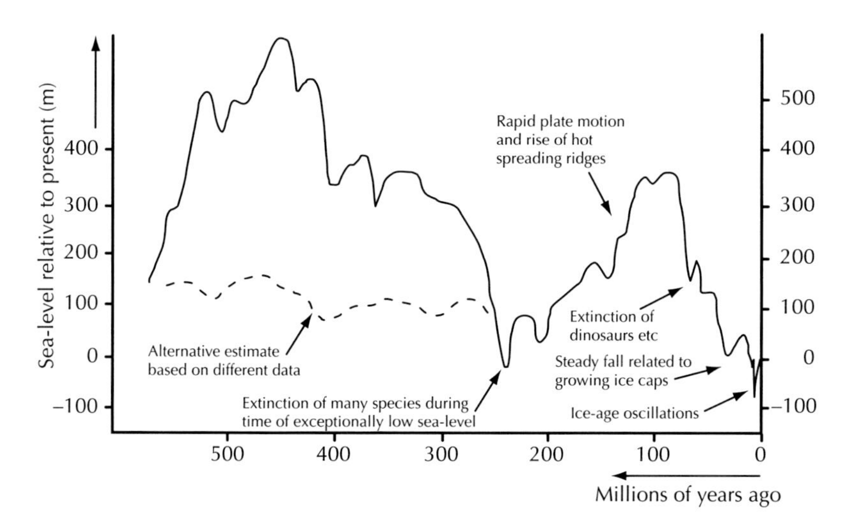

Diagram 21 Estimated world sea-level over the past 500 million years, relative to the present. World sea-level was much higher at the time of the dinosaurs, when rapid plate motion swelled the oceanic spreading ridges and displaced water onto the land.

global sea level was rising and falling throughout geologic history and that this was responsible for the sequences of sediment outbuilding seen on the seismic records. The pattern of global sea-level changes deduced from these seismic records was published by Exxon in the 1970s (diagram 21), although the data themselves remain confidential.

The approach used by the Exxon geologists to account for the shifting patterns of sediment advance and retreat seen on seismic records is called *sequence stratigraphy*. Units of sediment deposited during phases of progradation and later flooded are known as *sequences*. They typically overlie erosion surfaces (*sequence boundaries*) where the shelf has been exposed to erosion by a relative sea-level fall. This occurs when world sea-level falls more rapidly than the shelf is subsiding, or when tectonic effects cause the shelf to be uplifted faster than sea-level is rising; in either case, as explained in Chapter 3, accommodation space is reduced (diagram 22). Transport of sediments off the exposed shelf occurs at this time of low relative sea-level, called a *low-stand*. Deep marine low-stand deposits generally do not overlie a major erosion surface, since sea-level did not fall enough to expose the deep sea floor (diagram 23). Later in the development of the sequence, relative sea-level rises again. This happens because the shelf is subsiding faster than world sea-level is falling, or world sea-level is rising faster than the shelf is being uplifted; either way, accommodation space is increasing. Sediment

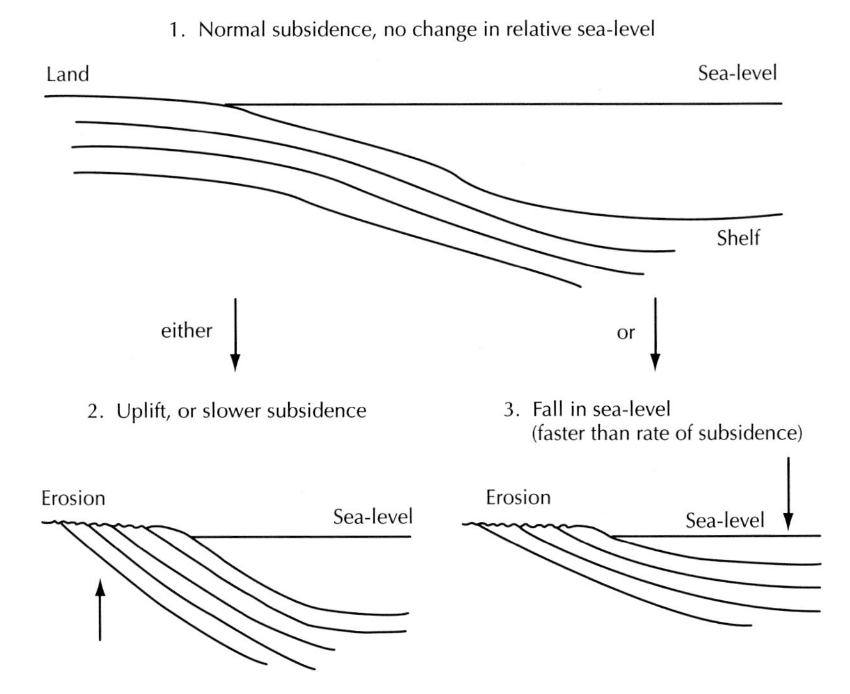

Diagram 22 Sketch to show how relative uplift of the land or a fall in world sea-level can reduce accommodation space and cause coastal erosion.

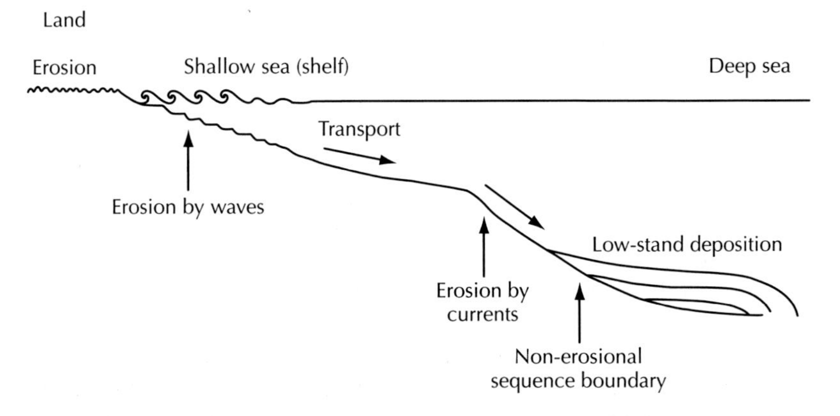

Diagram 23 A sea-level fall causes coastal erosion and waves affect sediments previously deposited in deep water. The eroded sediments are transported into the deep sea, where they rest on an uneroded surface. This episode of erosion and deep-sea deposition forms a sequence boundary.

is no longer transported into the deep sea but collects on the shelf. If sea-level rises too quickly for sediment supply to keep pace, the shelf is flooded. When sediment supply is fast enough to fill accommodation space as relative sea-level rises, the new space is filled with sediments and new land is built out onto the flooded shelf. These two phases are *transgressive* and *high-stand* deposition. The transgressive phase begins with a *transgressive surface,* formed as the sea begins to flood across the land. Wave action on beaches tends to erode a few metres of the underlying sediment at this time.

The greatest extent of flooding of the sea over the continental margin occurs at the end of the transgressive phase, known as the *maximum flooding surface.* This is often a time of reduced sediment transport out onto the shelf, because large areas of land are flooded and most of the sediment coming down rivers is trapped in estuaries and beaches. Stagnant muds or chalky limestones composed of marine shells tend to blanket the sea-floor at this time. High-stand deposition occurs as sea-level gradually stops rising and then begins to fall, and sediment supplied from the land progrades out onto the shelf. The later high-stand deposits are then eroded as sea-level falls more rapidly, beginning the deposition of a new sequence with its own low-stand, transgressive and high-stand deposits.

These definitions of the sediment deposition phases can be used to understand the process in any basin in terms of the balance between accommodation space and sediment supply. This is the main value of sequence stratigraphy. It is easier to apply it to areas where seismic records are available since the patterns of deposition are usually visible on seismic data. Without such evidence of progradation and flooding events, it is much more difficult to define sequence stratigraphic units. Very rarely, outcrops show patterns of sediment layers at a scale of hundreds of metres which can be interpreted in the same way as the seismic data. More commonly, studies of well records and outcrops in areas with seismic data are used to develop models of what sequence stratigraphic units should look like; these models are then applied to areas without seismic coverage. Examples are shown in diagram 24 – they must be used with extreme caution in interpreting new areas because no two basins are exactly alike in their patterns of sediment supply and changing accommodation space.

The idea that sequences due to global changes in sea-level were developed at the same time in different parts of the world is the basis of using sequence stratigraphy as a dating tool. If this idea is valid, the Exxon chart could be used to delimit the ages of sets of sequences identified in any marine sediments. Global sea-level also affects climatic humidity and the level of the water table so similar sequence stratigraphic dating might also be applied to non-marine basins filled with river, lake and desert deposits.

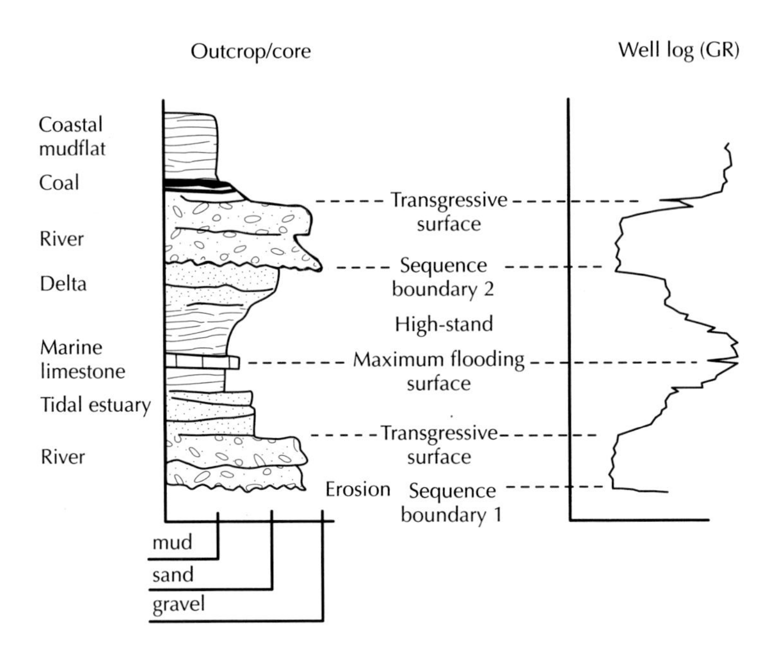

Seismic cross-section

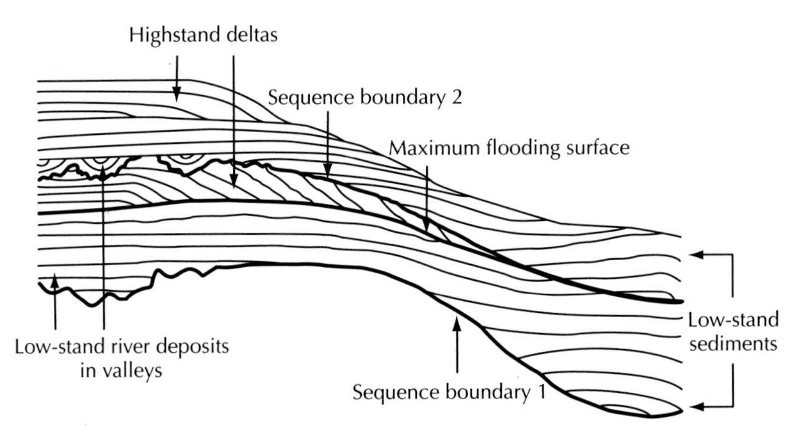

Diagram 24 Examples of sequence stratigraphic units seen at outcrop or in core, in well logs and on seismic sections. The seismic patterns provide the most reliable evidence of relative sea-level changes.

The duration and magnitude of the sea-level changes on the Exxon chart fall into four main categories. The longest-term ones, the *first-order cycles*, extend over hundreds of millions of years. They are thought to be caused by major long-term changes in the volume of the ocean basins resulting from episodes of slower and faster plate movement. At times of fast plate-tectonic spreading, the mid-ocean ridges are hotter and more inflated, displacing more ocean water onto the continents. *Second-order cycles* occur on 3–50 million years scales and are related to changes in the rate of subsidence of continental margins. *Third-order cycles*, 0.5–3 million years in duration, are responsible for the sequences seen most easily on seismic records. They come from short-term changes in accommodation space following variations in either tectonic subsidence or global sea-level (or both). *Fourth-order cycles* (lasting 0.1–0.5 million years) comprise the shortest-term episodes of sediment progradation and flooding, sometimes referred to as *parasequences*. During the Ice Age, these were generated by fluctuating ice volume at the poles, forcing world sea-level to rise and fall. The duration of these cycles was controlled by Milankovitch climate cycles, as described earlier. Changes in local sediment supply, for example by switching channel patterns in a delta, can produce cycles of sediment progradation at a similar scale without the need for world sea-level to change. Fourth-order sequences are best seen in areas with very rapid sediment accumulation; the Mississippi, Nile, Amazon and Niger deltas are all good examples. On most continental shelves, sediment accumulation is much slower and fourth-order sequences are consequently too thin to be seen on seismic data (which has a resolution of a few tens of metres). Most sequences observed at outcrop, with thicknesses of a few metres to tens of metres, are fourth-order sequences.

Recent work has questioned some of the early assumptions about the value of sequence stratigraphy in dating sediments. It is now known that some of the areas used to define third-order sequences in terms of global sea-level changes were undergoing rapid tectonic uplift or subsidence when the sequences were deposited. This makes it difficult to unravel the competing effects of tectonic changes in accommodation space, sediment supply from uplifted areas and global sea-level. It is also thought that widespread changes in tectonic subsidence can produce near-synchronous third-order changes in accommodation space in widely separated basins, which may appear to be global if a limited database is used. Proving that fourth-order sequences developed at the same time across the globe is currently impossible for most pre-Ice Age sediments because the accuracy of independent dating methods (mainly biostratigraphy and radiometric dating) is not good enough at the short time scales involved. Moreover, it is difficult to find a plausible mechanism for rapid, short-term global sea-level change before the polar ice caps developed about 30 million years ago. Despite these reservations, the developing science of sequence stratigraphy has encouraged geologists to think carefully about how sediment deposition responds to changes in global sea-level, subsidence and sediment supply, and to investigate such changes from limited outcrop or well data.

Other types of event stratigraphy

These relative dating methods – biostratigraphy, magnetic stratigraphy, stable isotope stratigraphy and sequence stratigraphy – all rely on identifying very widespread or global events in the sedimentary record. Within an individual basin, a local stratigraphy can be developed by reference to basin-wide events which may include falls of volcanic ash, major storms or episodes of tectonic uplift.

More widespread events, outside these categories, are also worth a mention. They include major asteroid or comet impacts which may be recorded by their effects on biostratigraphy: one famous example is the demise of the dinosaurs and a host of other organisms at the end of the Cretaceous Period, which is believed to have been caused by a momentous comet or asteroid impact. Some episodes of volcanism, such as the periodic eruption of "supervolcanoes" as seen at Yellowstone National Park, have a major influence on sedimentation and biostratigraphy.

The largest outpourings of volcanic lavas in the last few hundred million years occurred in Siberia at the end of the Permian Period and in India at the end of the Cretaceous. Both are associated with mass extinctions of marine and land organisms, as well as with evidence of large impact events although the end-Cretaceous impact is the more clearly recorded. Since the Indian eruptions began about 1 million years before the comet or asteroid hit at the end of the Cretaceous, they cannot have been caused by the impact. However, most of the lavas appear to have been erupted in a brief episode of about the same age as the impact in Mexico, suggesting that the increased volcanic activity in India might somehow be linked to Earth's collision with a comet (perhaps due to a second impact in India analogous to the multiple impacts of Comet Shoemaker-Levy on Jupiter).

Global events also embrace the so-called *oceanic anoxic events*, times of severely reduced oxygen content in the world ocean which took place during the early Jurassic and mid-Cretaceous Periods. They appear to have followed from a number of inter-related factors: warm global climate, increased volcanic activity, restricted circulation of the world ocean and high global sea-level. The dominant factor in all these events seems to have been plate tectonics. Very active plate tectonic spreading would have caused the mid-ocean ridges to expand, displacing the ocean onto the continents.

Shallow seas spread across many of the continents. At the same time, high levels of carbon dioxide, erupted from volcanoes during plate tectonic activity, caused global warming because of the greenhouse effect. Abundant organisms thrived in the warm, shallow seas, and when they died their decaying remains removed oxygen from the sea water. Ocean circulation was sluggish, possibly due to stratification of the ocean into a dense, saline, low-oxygen layer and a less saline upper layer, and this may have been the factor preventing replenishment of oxygen from the atmosphere and hence establishing a period of global oceanic anoxia.

Eventually, plate tectonic activity slowed down again, adding less carbon dioxide to the atmosphere. Carbon was removed from the atmosphere/ocean system in the form of dead and buried organisms so global carbon dioxide levels fell. As sea levels fell and the Earth cooled, sea-levels everywhere fell, fewer organisms lived and died in the seas, and the normal oxygen levels of the oceans were slowly re-established. It is interesting to note that the climatic and oceanic models built up in studying such events might apply to a future greenhouse world were current global warming to continue.

Chemical stratigraphy

A new method being used by the oil industry for relative dating of sediments is the establishment of a local chemical stratigraphy. From the bulk chemical composition of cuttings samples collected from a well, the variations in the proportions of the various chemical elements can be plotted against depth to show up changes in sediment composition as different rock types were eroded to supply sand and silt to the basin. The varying patterns of sediment supply can be recorded in much the same way within small areas, although at any one time widely separated parts of the same basin will receive different mixes of sediment. The technique is therefore used mainly for relative dating within individual oil fields across distances of just a few kilometres. The possibilities of caving and post-depositional chemical alteration of sediments always have to be taken into account and the technique is used mainly in sediments containing no age-diagnostic fossils.

Cyclostratigraphy

For over 100 years, geologists have realised that the cyclic changes in the Earth's orbital parameters might be recognisable in sediment sequences and, indeed, careful recent studies of Ice Age sediments have proved that the various Milankovitch cycles are recorded in deep-sea sediments. We are now beginning to recognise the approximately 20,000, 40,000 and 100,000 year Milankovitch climate cycles in pre-Ice Age sediments, with the exciting prospect of dating events in Earth history with a precision hitherto only dreamed of.

The first studies of this type relied on field observations of repeated changes in rock types and depositional environments on the scale of a few metres which were regarded as representing time scales of around 100,000 years. Sometimes "bundling" of sets of five thin cycles into longer-term units were observed, suggesting a record of five 20,000-year precession cycles in each 100,000-year eccentricity cycle. Cycles of advance and retreat of deltas, changes in the flow power of rivers, differences in lake depth and chemistry, and variations in oceanic carbonate production and deep sea current strengths could all be documented in great detail but the precise timing of these changes was always very difficult to determine. No independent dating method exists that can measure the duration of such cycles with the accuracy needed to prove a link to Milankovitch cyclicity.

A detailed understanding of the Ice Age climate has developed from oxygen isotope compositions of deep sea cores and ice cores leading to an understanding of how, over the last 2 million years, sedimentary systems ranging from deep sea floors to deltas and lakes have reacted to proven changes in climate and sea-level, driven by Milankovitch cycles. The big challenge is to show that similar cycles of sediment deposition in earlier times were also geared to Milankovitch cycles – that would open the possibility of dating events in Earth history with an accuracy of better than 100,000 years.

The data which showed that oxygen isotope levels in Ice Age cores were driven by Milankovitch cycles were investigated by *Fourier Transform Analysis* which detects cyclic variations in the data and shows how much of the total signal is contributed by the detected cyclicity. The procedure only works when the sediments are deposited at an almost constant rate because it detects only near-perfect "sine-wave" cyclicity. All stages in the cycle, changing gradually from one extreme to the other, must be present in the record. If the record of changing Ice Age climate were interrupted by breaks in sediment deposition (such as periods of sea-floor erosion), it would be impossible to detect cyclic patterns in what remained. Luckily, the deep-sea sediments were produced by an almost constant rain of microscopic shells onto the sea-floor which accumulated as limestone sediments at an equally constant rate. But remember that most sedimentary sequences are deposited episodically, with rare storms and gravity-driven flows contributing most of the preserved sediment. Long periods of time go unrecorded in such sediments, so Fourier Transform Analysis would detect no cyclicity in any climatic signals recorded by such episodic sedimentation.

Simple sedimentary successions comprising two main rock types can be analysed by *Walsh Power Spectral Analysis*. Provided the two rock types represent the products of a climate-driven variation, this analysis can demonstrate Milankovitch cycle periods in long, continuously-exposed sections.

Unfortunately, most rock sequences are too complex but there has been some success in tracing the existence of Milankovitch cycles in deep sea limestone-mudstone sequences.

A more recent statistical technique to be used is *Maximum Entropy Spectral Analysis (MESA)*, a method of describing any set of data as accurately as possible in terms of superimposed cycles. It was developed for automated voice recognition, since each person's voice contains distinctive, characteristic vibrations determined by the shape of their voice box, a pattern overprinted onto the sound of the voice. If the voice is raised in pitch, a different set of sound wavelengths is produced but with the same overprint from the voice box. By analysing the sound wave cycles present in the voice, it is possible to identify each person's characteristic voice-box overprint and so recognise their voice.

The mathematics behind this analysis has now been used to analyse the signals contained in well logs (Chapter 3), a set of data recording rock properties which vary as a function of depth just as a voice is a set of sound waves varying in wavelength. By combining a set of superimposed cycles, or waveforms, to mimic the real patterns in the log data, MESA will attempt to reproduce the log as accurately as possible. Real log data is not composed of perfect cyclic waveforms, just as real voices are not perfect sets of pure musical notes. Despite the background noise, MESA will very usefully reveal any cyclic variations which *are* present in the log.

The most useful well log for the detection of climatic cycles is the gamma-ray log, which mainly responds to variations in the proportion of sand and mud in the sediments. This is because a typical fourth-order sequence consists of sands deposited in higher-energy conditions building out over deeper-water muds due to progradation, then being flooded and finally covered with quieter, deeper-water muds. Repeated episodes result in cyclic changes from sandier to muddier sediment recorded by parallel cyclic changes from lower to higher gamma-ray readings in borehole logs (diagram 25). In effect, MESA measures the average thickness of these fourth-order "parasequence" cycles (diagram 26). Note that on all of the cyclicity diagrams, longer cycle wavelengths are shown to the left and the wavelength scale is logarithmic.

The analysis is performed on a few tens of metres thickness of the sediments, providing a measure of the cycle thicknesses contained within that part of the sediment pile, which is called the *window of analysis*. Next, a section of the same thickness is analysed beginning slightly further up the well and mostly overlapping the last window. A similar set of cycle thicknesses is measured. The analysis proceeds stepwise up the whole log, through hundreds of metres of sediment, in what is called a *sliding window analysis*. The result shows any gradual changes in the thickness of the cycles present in the log (diagram 27). The superimposed sets of cyclicity patterns are more easily seen in colour

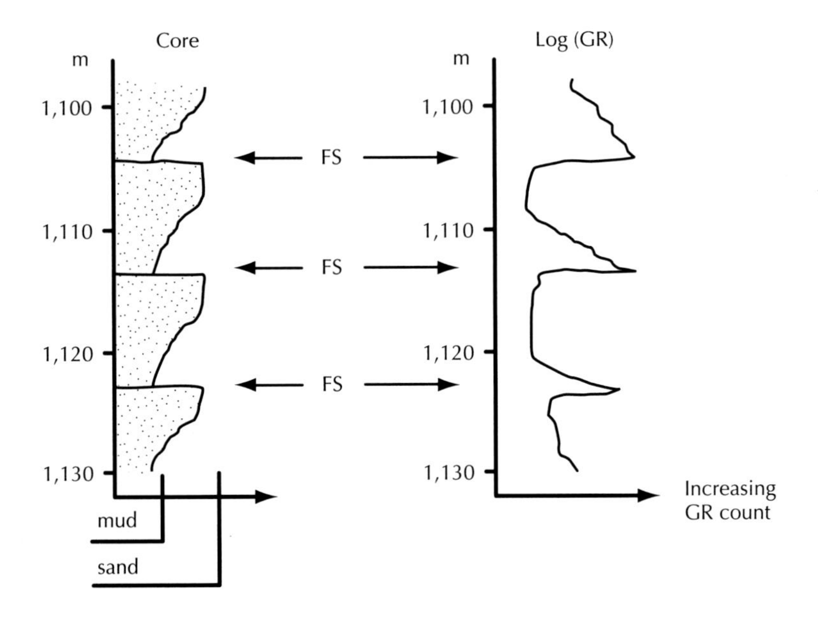
a) Gamma-ray log record of parasequence cycles of flooding and sediment progradation
Core
Log (GR)
m
m
1,100
1,110
1,120
1,130
1,100
1,110
1,120
1,130
FS
FS
FS
mud
sand
Increasing GR count

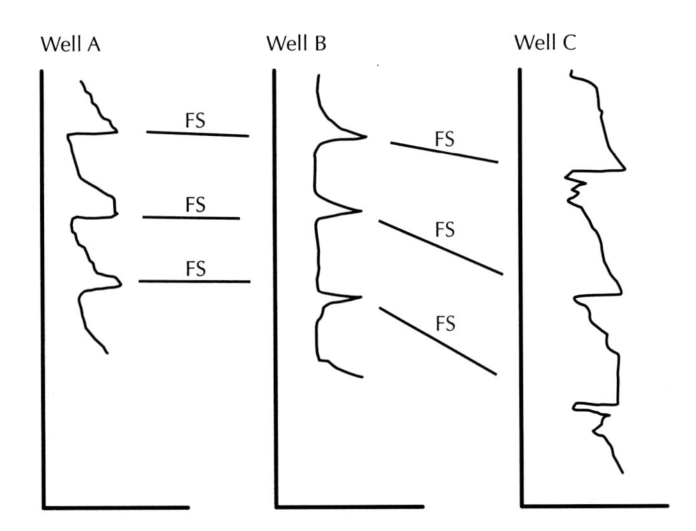
b) Gamma-ray logs in adjacent wells record the same set of parasequences
Well A
Well B
Well C
FS
FS
FS
FS
FS
FS

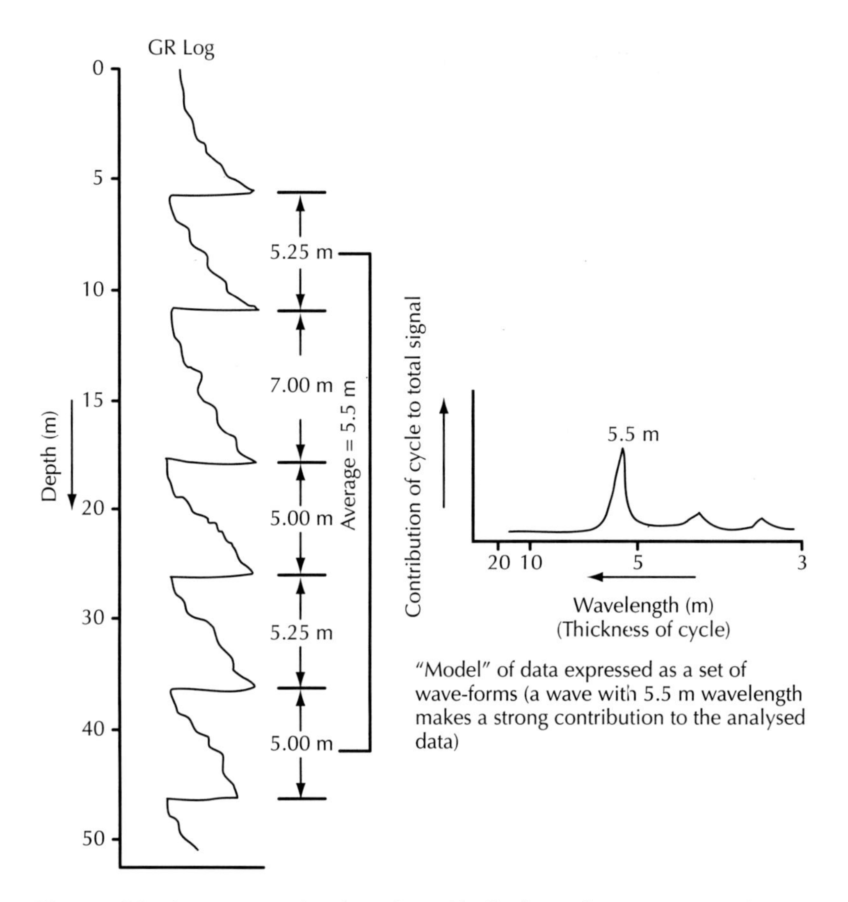

Diagram 26 A gamma-ray log through an idealised set of parasequences shows a clear cyclicity but with varying cycle thickness. Maximum entropy analysis shows the average cycle thickness (in this case 5.5 m), by determining that a 5.5 m-long wave-form would simulate most of the variation in the gamma-ray log.

Diagram 25 a) Small-scale variations in sand supply to a coastal area record periods of sea-level rise when muddy deposits formed, known as flooding surfaces (FS). The flooded areas were filled in with sand between each flooding event, producing cyclic parasequences. These are readily detected on gamma-ray logs.

b) Gamma-ray logs from adjacent wells usually record the same set of parasequences bounded by flooding surfaces (FS). The parasequences are probably all of similar ages in closely-spaced wells.

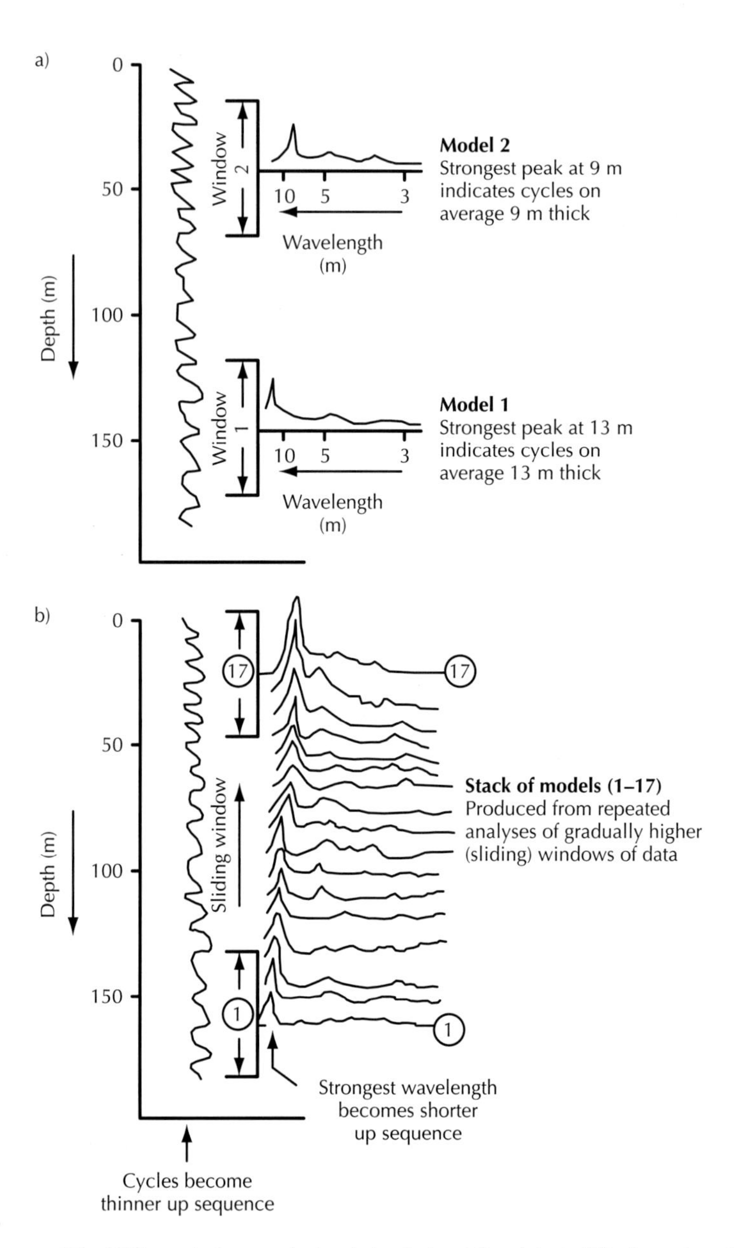

Diagram 27 Sliding window analysis. a) Analysis of the data in Windows 1 and 2 generates different results, because of thicker cycles in the lower part of the log. b) A series of analyses, performed on progressively higher windows up the log, generates slightly different results which show how the cycle thickness changes up the log.

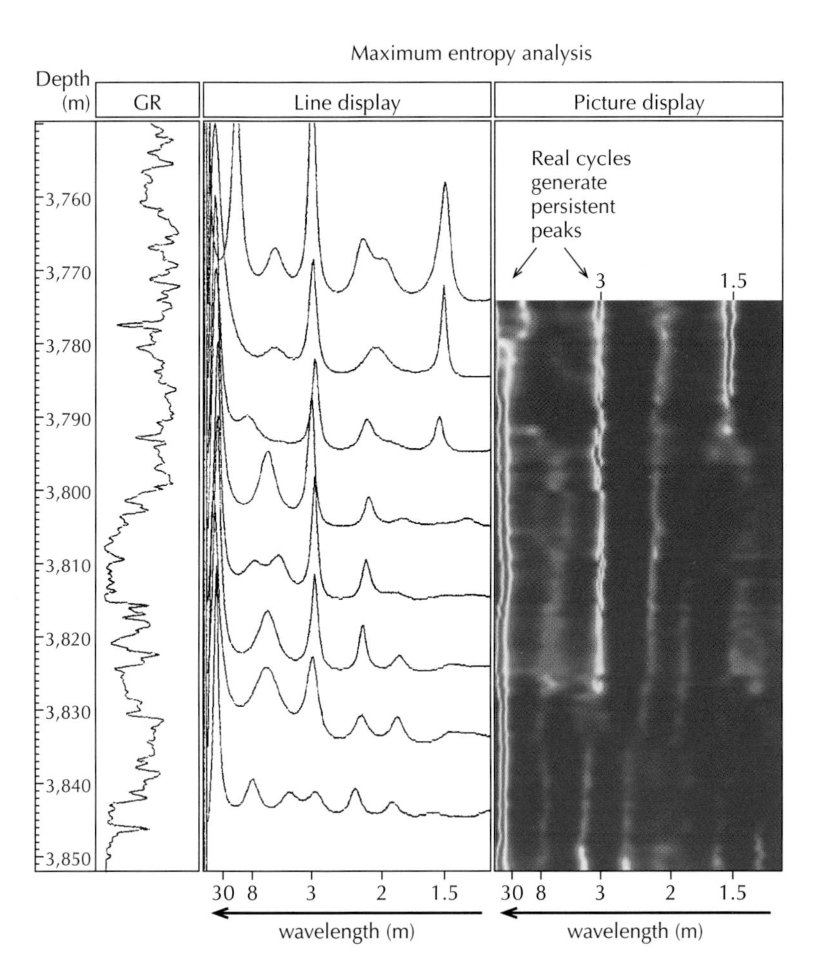

Diagram 28 Cyclicity analysis of a gamma ray log from an oil well in the North Sea shows persistent cyclicity of 30 m and 3 m thickness, displayed as line graphs and as a picture. Only a few of the line graphs produced by the analysis are shown, for clarity. Peaks on the line graphs are shown as bright areas in the picture display.

rather than as a set of line graphs (diagram 28). Strong cycle wavelengths are usually shown in red on full colour displays, or in white on monochrome displays.

Sliding window analysis performs a second useful task. MESA is a modelling procedure which mimics any set of data (no matter how random) as a set of cyclic waveforms. Analysis of each window of data is *bound* to result in a set of cycles which most closely approximate the patterns contained in the data, because that is what the technique is programmed to deliver. But if the log data were truly random, analysis of each separate window would produce a different (highly inaccurate) model, expressed as a set of distinct cycles and no consistent model would result. Sliding window analysis of real log data generally shows strong cyclicity which persists for many tens of metres up the log, indicating that the log contains *real* cyclicity which is consistently recognised by the analysis (diagram 28).

There is a third reason for using sliding window analysis. In practice, two sets of analyses are run on each log using two different sliding window sizes. The first may analyse, say, a 30 m thickness of log data in each window while the second may analyse 50 m. This allows us to identify positions on the log data where cyclic patterns change abruptly (diagram 29). As the window of analysis moves up the log data, a longer window will pick up any change in log patterns before a shorter window. Conversely, the top of a log pattern is recorded sooner by short-window analysis while it still affects long window analysis. Comparison of short and long window analyses shows us which log patterns were generated by which sections of log data.

This analysis is performed to define an objective, machine-based measurement of the thickness of repeated cycles of sediment deposition. Gamma-ray logs identify cycles of sandier and muddier sediment deposition. Because they respond to the muddiness of the sediments in much the same way, neutron and density logs from the same rock sequence show similar cyclicity to gamma-ray logs, which must therefore be due to changes in the original sediment composition, from sand to mud and back again. The persistent presence of such cycles leads inevitably to questions about their origins.

The thickness of the repeated sediment cycles is shown by the wavelength of the detected cyclicity, and it is common to record several superimposed cycles of different thickness in any part of a log. Diagram 30 shows a log with several co-existing cycles of different thickness. What does this mean?

In most cases, subsidence of a sedimentary basin is gradual and continuous at the time scale of fourth-order sequences (10,000–100,000 years). As defined by radiometric dating and biostratigraphy, subsidence rates of about 100 m per million years are typical of most extensional basins. MESA analysis of well logs is typically able to image cycles with thicknesses of 2–30 m. In most basins, these thicknesses represent 20,000–300,000 years, but we can only prove this if independent data are available to confirm the age of two or more surfaces within the sedimentary sequence. Biostratigraphic or isotope evidence is needed to prove the approximate time-span of a known thickness of sediments and hence the approximate rate of sediment

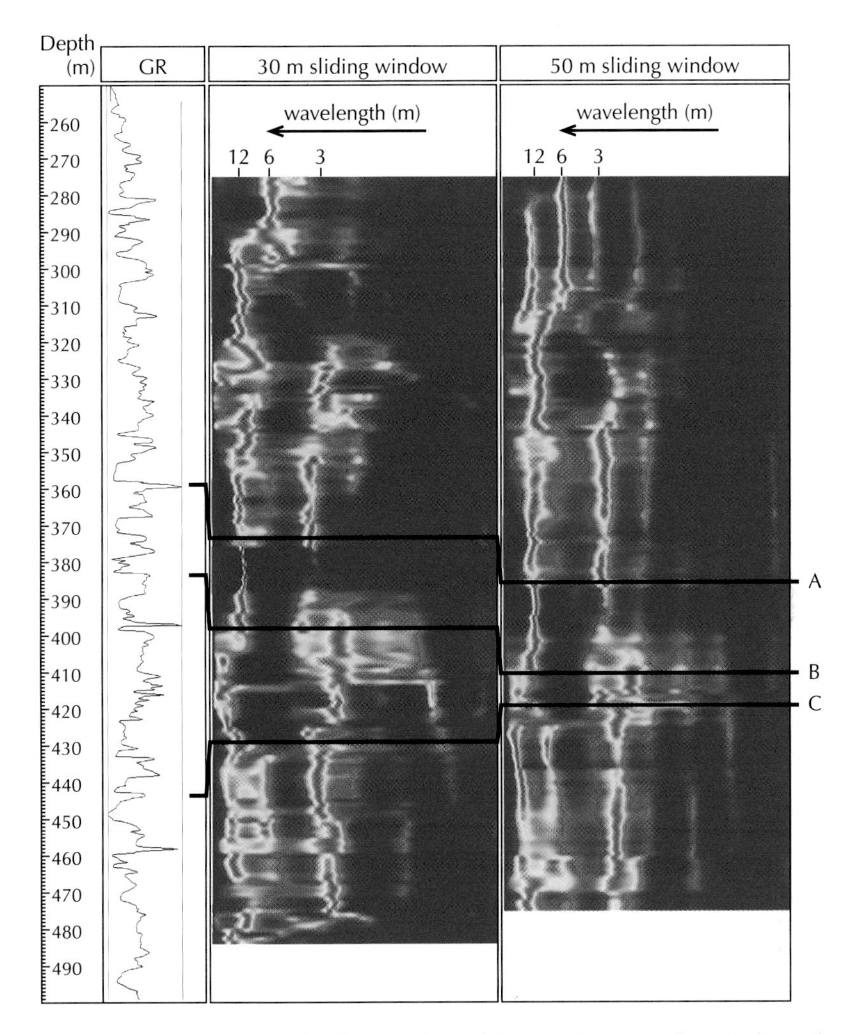

Diagram 29 Comparison of cyclicity analysis of the same log, using data windows of 30 m and 50 m length. Changes in the cyclicity patterns at A, B and C can be traced up (or down, as appropriate) from one set of results to the next, and then back to the log data.

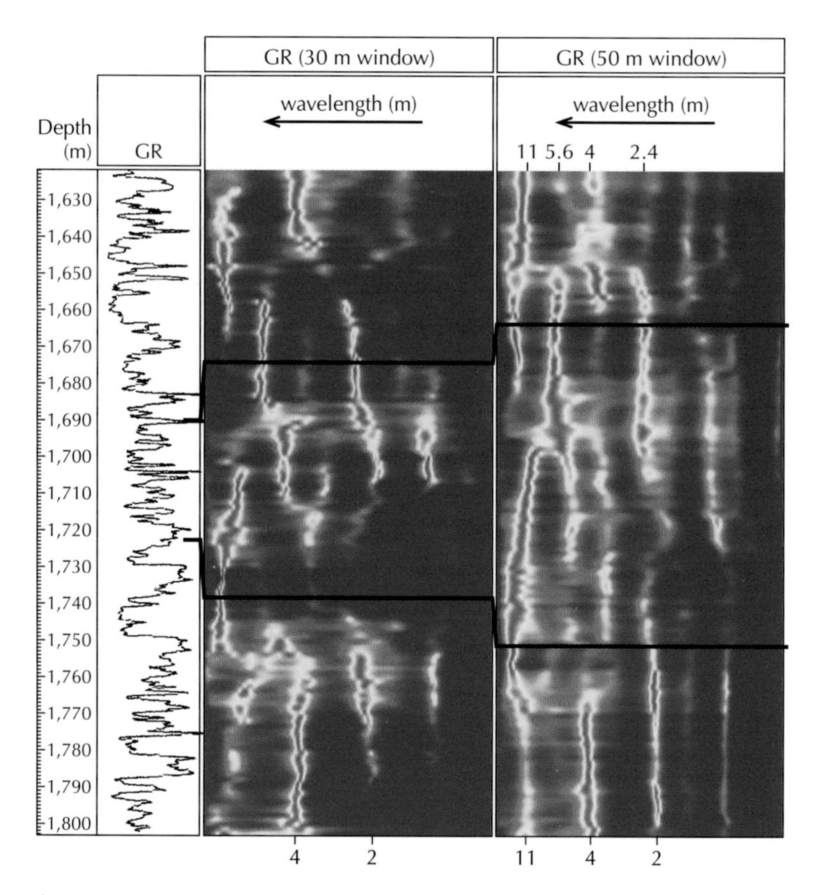

Diagram 30 Log with several co-existing cycles of different wavelengths. The main cycles present have thicknesses of about 11 m, 4 m and 2 m in the upper and lower parts of the log, with cycles of 5.6 m and 2.4 m thickness present in the middle part.

deposition in the basin. It is then possible to estimate the time represented by the cyclic thicknesses of sediment detected by the analysis.

The final step is to identify possible Milankovitch cyclicity in the log. If we know from independent dating that the sediments were deposited at a rate of about 10 cm per 1,000 years, and the analysis shows strong cyclicity with a thickness of, say, 11 m, one can postulate that a 100,000-year eccentricity cycle is being recorded. The additional presence of cycles with thicknesses of 4 m and 2 m (as in diagram 30) would probably represent the 41,000 year obliquity and 21,000 year precession cycles, so supporting (but not proving) the interpretation that the 11 m cycle represents the 100,000 year eccentricity cycle. Then, if the part of the log which contains 11 m cycles is 110 m thick, one can estimate that it represents a total of 1 million years (assuming that there are no major missing sections within the log, perhaps because of faulting or erosion).

Because of the lack of precise independent age data, it is very difficult to prove that the cyclic deposits in a well really represent Milankovitch cycles. Probably the most convincing way of doing so is to analyse a large number of wells within a small area. If one can be sure that a certain sequence of sediments, present in all the wells, records the same length of geological time, it is possible to test the Milankovitch cyclicity approach (diagram 31). The analysis should indicate that the sequence records the same amount of time in each well. Furthermore, this should be the case no matter which Milankovitch cycles are identified in each well – in one, the 400,000 years recorded by the sequence may be represented by four 100,000-year eccentricity cycles while in another we may find ten

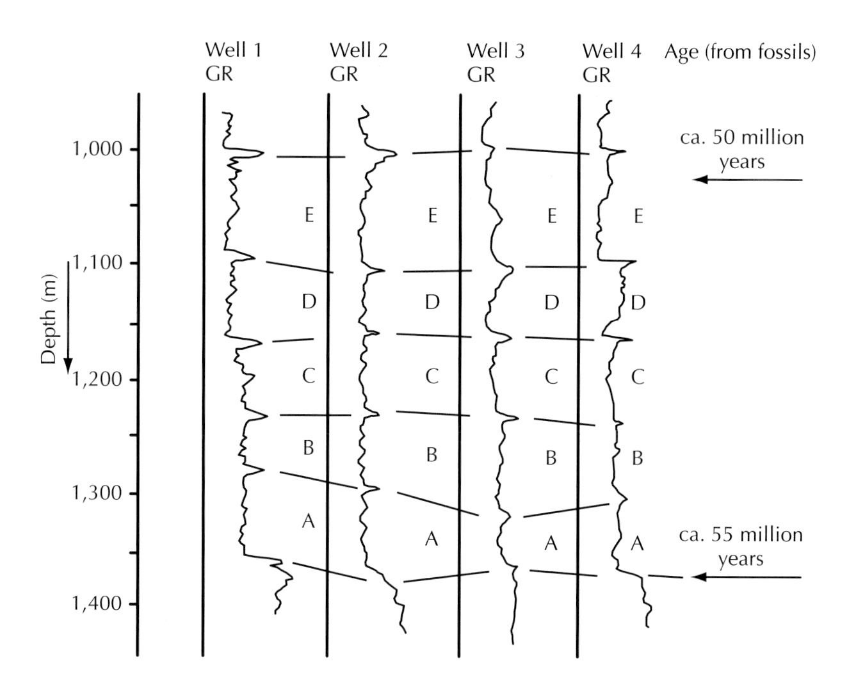

Diagram 31 Sketch showing gamma-ray logs from four adjacent wells. Visual correlation of the logs, as shown, suggests that units A–E can be traced across the area. Age determinations indicate that the total sequence spans about 5 million years. One would expect each of the five units to represent about 1 million years, i.e. each should contain ten 100,000-year cycles (assuming the sequence is essentially complete).

41,000-year obliquity cycles. A convincing demonstration of this kind has now been achieved in a study of 19 wells drilled in an area 40 km across in southern Yemen. That study, described in Chapter 6 as a detailed case history, provides very precise timing of events over 120 million years ago when dinosaurs roamed swamps and river plains in an area now covered by desert.

6

Snapshots from the past

In this section, we will start with a study of Nepal and Pakistan (which involved me directly) to show how geologists work in the field and to explain a little about why, about 8 million years ago, our far-distant ancestors began leaving their forest homes to adopt a nomadic hunter–gatherer lifestyle. Then we will move on to southern Yemen, to examine oil-industry data which shows quite clearly that what is now a desert was once a complex of rivers, tidal creeks and mud-flats, being flooded by the sea at the time of the dinosaurs, 120 million years ago. The geological history of the area can be worked out in great detail from climatic rhythms encoded in the rocks. The episodic flooding of the flat-lying coast by the rising sea was mainly a response to regular changes in climate driven by 100,000-year and 40,000-year Milankovitch cycles.

These thumbnail sketches of oil-industry geology also illustrate some of the recent improvements in our understanding of the way the Earth works. In recent years this has been greatly aided by ever increasing computer power and the interaction of people from different backgrounds via the Internet. Such rapid advances in our understanding seem likely to continue. There is still a vast amount to be known about our planet; it can sometimes be a frighteningly unstable place and so far we have been relatively lucky. But the more we learn about our environment, the more able we will be to deal with problems as it changes.

The rise of the Himalayas and the development of Man: Nepal and Pakistan, 1988–1996

In collaboration with the University of South Carolina, I worked in the spring and autumn of 1988 as the director of a study of southern Nepal, on behalf of an international oil company. Our team included experts from America, Australia, Britain and Nepal. Our job was to describe and measure the thickness of the pile of sediment which was eroded from the Himalayas as they rose. The mountains were rising at least 40 million years ago, shedding mud, sand and gravel into a shallow sea separating Asia from what was then the north coast of India. Huge amounts of sediment were transported southwards from the rising mountains by south-flowing rivers. The over-supply of sediment soon filled the shallow sea, creating a vast swampy plain in what is now southern Nepal, crossed by major rivers. As the land in front of the growing pile of thrust sheets on the southern side of the Himalayas subsided during perhaps 10 million years, a foreland basin developed and filled with thousands of metres of sediment (see diagram 32). Finally, about a million years ago, the whole pile of wet, poorly compacted sediment was buckled up and deformed into a series of rucks and slices as the Himalayas pushed southwards, forming the still-rising Siwalik Hills. Continuing earth movements in these hills caused a major earthquake which we felt on arrival in Kathmandu, and destroyed many of the bridges in our field area.

The oil company sponsors wanted to know how far they would have to drill through this young sediment before reaching their objectives in the

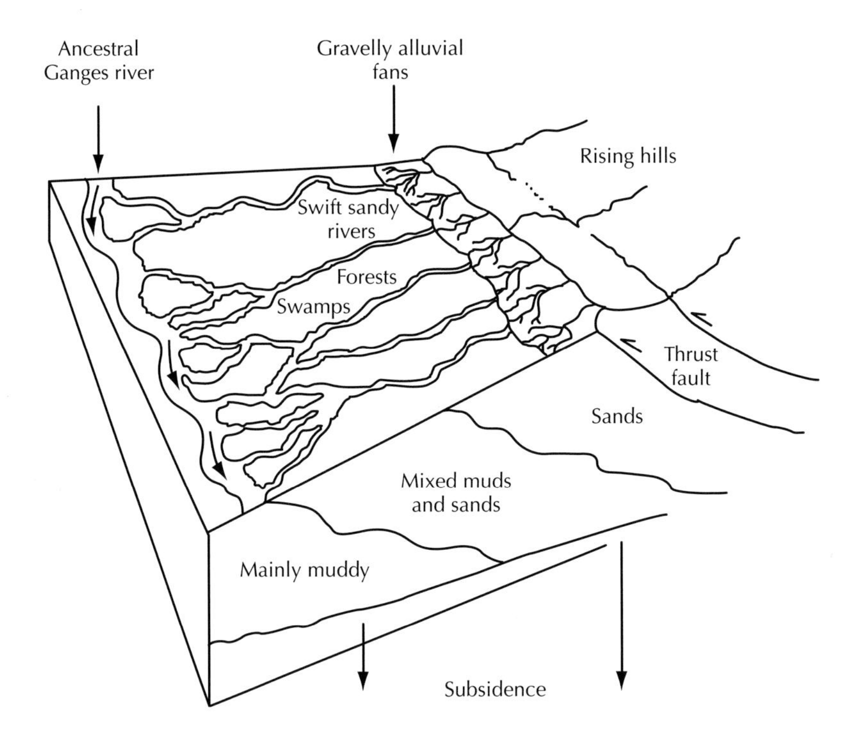

Diagram 32 Reconstruction of environments in southern Nepal about 10 million years ago. As the Himalayas pushed southwards, belts of muddy and sandy sediment advanced out into the subsiding basin, transported by numerous rivers which fed into the ancient equivalent of today's Ganges River.

older, marine sediments which formed when the mountains were first rising. Our main task was thus to work out the thickness at a number of potential drill-sites in southern Nepal. That meant piecing together the sequence of sediments now exposed as slices of rock, each several kilometres thick, which have been piled up on top of each other, tilted and then dissected by deep gorges.

Our first requirement was to find a major north-south gorge cutting across the pile of thrust-slices and describe the sequence of rocks exposed there. We found the ideal locality at Surai Khola, where a deep gorge cuts southwards through the Siwalik Hills. Using methods described in Chapter 3, we logged the exposed section, over 4000 m thickness of sediment (diagram 33).

Meanwhile, we worked out how each thrust slice had been pushed southwards as the Himalayas rose during the convergence of the India and Asian Plates, and mapped the whole region by walking through the hills and gorges to report variations in the stratigraphy and structure to our clients. From the Surai Khola section, 128 orientated samples of mudstones were collected for magnetic dating back in the laboratory. These mudstones had been deposited on the flood-plains of the southward-flowing rivers and contained iron minerals which crystallised as the rocks were buried; those crystals aligned along the Earth's magnetic field at the time of burial. Analysis showed a pattern of repeated reversals of the Earth's magnetic field over the last 10 million years, a pattern which matched precisely with records from the ocean floor to give a fairly confident age assessment for the Surai Khola sequence. The match was made possible by establishing a few known ages using dozens of fossils, many of which were of the same species as fossil mammals found in precisely-dated sediments in Pakistan during earlier studies by members of our team.

Our survey showed that in many ways the rocks in southern Nepal resembled those in northern Pakistan: of similar age and also deposited by rivers flowing away from the rising Himalayas. The Pakistani samples had been dated both by magnetic stratigraphy and by fossils. The sediments in Nepal and Pakistan all showed evidence of ancient soil formation on well-drained parts of the river flood-plains. The soils had been formed when rain-water percolated through the vegetated flood-plains to the water-table, where dissolved carbonates were precipitated as lumps and ledges of limestone. The ledges are now tilted along with the rest of the rocks, proving that they formed as the sediments were deposited on flat-lying flood-plains, rather than being caused by more recent weathering effects. The limestone nodules are composed of calcium carbonate containing stable carbon isotopes, in concentrations suggesting the types of plants growing on the flood-plains at that

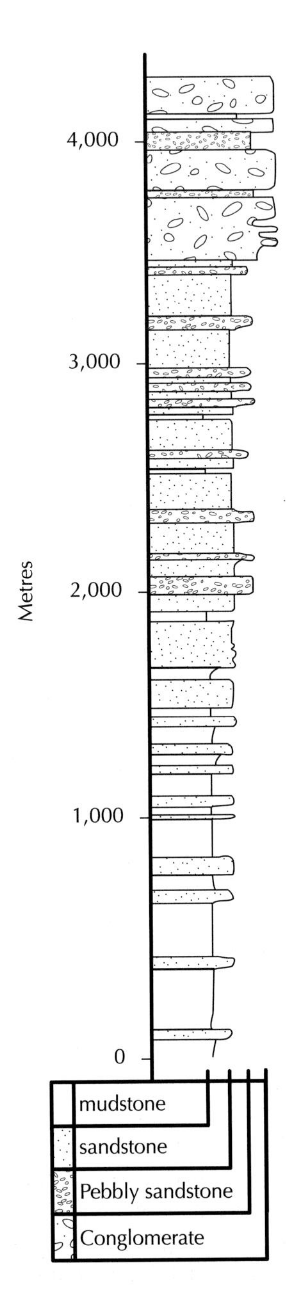

Diagram 33 Simplified version of sedimentary log through the 4000 m-thick sequence exposed in the Surai Khola gorge, southern Nepal.

time. The isotope contents have remained essentially unaltered since the soils first formed.

The soil carbonates had already been shown to record a dramatic change in the vegetation in northern Pakistan during the late Miocene epoch. Forests and shrublands dominated the flood-plains in damp, humid conditions until about 7.6 million years ago, after which the trees and shrubs were gradually replaced by grasses. This change was complete by about 5 million years ago and was thought to be linked in some way to the increasing effects of the Asian monsoon, which had already been detected by studies of microscopic organisms in the Arabian Sea. These record increasing monsoonal effects after about 7 or 8 million years ago. The monsoon was therefore believed to have intensified some 8 million years ago, presumably because of increased uplift of the Himalayas (or more precisely, increased elevation of the Tibetan plateau). Monsoon winds are generated by the development of low-pressure weather systems over the relatively warm, elevated Tibetan plateau (averaging 5 km above sea-level). The rising warm air draws in cooler air from the Arabian Sea and Indian Ocean over the Indian sub-continent, causing intense rainfall during the hottest part of the year. The reverse occurs in winter, blowing dry air outwards from the Himalayas to bring long winter droughts. Grasses are better adapted than shrubs and trees to long droughts and a short, hot growing season, and also recover better from bush-fires started by the lightning-strikes which herald the coming of the monsoon rains. It was reasonable to argue that the soil samples recorded the increasingly intense effects of the rising Himalayas on the Asian climate.

The soil nodules found in Nepal resembled those from Pakistan; one might expect them to record similar compositional changes if the nearby Himalayas were influencing the regional climate and vegetation patterns.

We used our records of magnetic reversals and fossil mammal records to date the Surai Khola section accurately and to establish the age of our soil samples. The intention was to establish the precise pattern of climate change in Nepal. We could then sample less complete, poorly-dated exposures elsewhere, using the Surai Khola section as a basis for fitting all the other data from the region into a well-defined sequence. This would allow us to piece together the thicknesses of rock exposed in separate areas, to build up the overall thickness information required by our clients.

The ancient soil samples collected from Surai Khola showed that trees and shrubs dominated the swampy flood-plains of southern Nepal until about 7 million years ago, to be gradually replaced by grasses over the following 2 million years. We could therefore define a "pre-transition" stage (over 7 million years old), a "transition" stage (7–5 million years old), and a "post-transition" stage (less than 5 million years old). Samples of soil carbonate from other areas across the region fitted into this pattern, a conclusion supported by further magnetic stratigraphy in some of the deep, rocky gorges in south-eastern Nepal.

The study helped us to work out where all our exposures of sediments fitted in to the general 10 million year-long sequence and gave the client oil company some advantage over its competitors in the ongoing search for oil in Nepal. It also showed that similar climatic and vegetational changes had occurred in Nepal and Pakistan, although the changes apparently happened about half a million years earlier in Pakistan. The regional extent of the vegetation changes strongly supported the idea that they recorded a widespread climate change. In 1996, a different oil company commissioned another project, this time in western Pakistan. This provided further information on the patterns of climate change over the last 10 million years across southwest Asia, allowing the regional picture to be seen. This can be fitted into our current understanding of the development of our early ancestors, which is summarised below.

It seems that the Tibetan Plateau has risen in elevation by about 5 km during the past 10 million years. By 8 million years ago, the plateau elevation was sufficient to cause (or enhance) the Asian Monsoon. At the same time, erosion and weathering of rocks in the rising High Himalayas and other mountain belts worldwide was removing carbon dioxide from the atmosphere (as dissolved bicarbonate salts). The drop in the concentration of carbon dioxide was exacerbated by a reduction in volcanic activity at the Earth's mid-ocean spreading ridges. The resulting atmospheric changes favoured grasses over trees and shrubs, and the warm growing season and long winter droughts caused by the Asian monsoon further tipped the balance in favour of the grasslands.

Grasslands also expanded in other parts of the world around this time, partly because of the global carbon-dioxide reductions and partly because mountains such as the Rockies were rising to influence local climate in a similar way to the Himalayas. It is likely that the main factor favouring grasses over trees and shrubs was their resistance to unstable, seasonal climates, giving them an advantage over rain-forest plants in the sub-tropics as global climate became more changeable with the gradual onset of the Ice Age.

Forests retreated worldwide as grasslands spread across many subtropical ("savannah") regions. In South and East Africa, early man-like hominid apes were dwelling in the retreating forests. Their habitats became crowded, forcing them to compete with larger apes or other specialists. Some of the more adventurous or adaptable types left the forests to occupy the expanding grassland areas where they had to stand up to look out for predators and spot their kills. A few million years later, Man may have started using weapons to drive scavengers from these kills – no mean feat when facing the giant hyenas and wolves of the time. People became nomadic, developing language, marching songs and perhaps (like the Aborigines) rhythm maps to chart their progress. They migrated, following game herds or the rising sun or their favourite stars, and survived by gathering wild fruits and vegetables when they couldn't get meat. To get more meat they developed better weapons and other tools. Many of the early migrations were along the coasts of Africa and Asia, where tools were invented for opening shellfish and butchering the carcasses of stranded sea-creatures.

Meanwhile, the climate was still cooling. By 2 million years ago, the stage was set for the start of the great polar glaciations of the current Ice Age. Global circulation of oceanic currents, which had kept the polar regions relatively warm and the climate fairly stable, was being disrupted by plate tectonic forces; eventually the system became critically unstable. Triggered by Milankovitch climate change and enhanced by natural feed-back effects ("climatic snowballing"), several dramatic falls in global temperature caused the spread of huge ice sheets into much of northern Eurasia and America. The harsh conditions prevailing in habitable valleys on the edges of the ice sheets were home to early Man who harnessed fire in order to survive and began to make warm clothes using newly developed tools. The most adaptable groups spread rapidly every time the ice sheets retreated, prospering in the temporary "Gardens of Eden" which flourished as global temperatures rose. During the wet phases associated with melting of the ice caps others migrated from central Africa into the Sahara, only to be forced northwards into Eurasia at the start of each new dry phase. Dramatic climate changes included natural warming and cooling events taking place over the space of a few decades – changes five times larger than the global warming commonly blamed on industrial activities in the last 150 years. These are mainly due to rapid changes in the behaviour of ocean currents.

World sea-level has risen and fallen by 50–150 metres during repeated phases of ice-cap melting and then re-growth over the past 1.8 million years; the most recent ("Flandrian") flooding phase has drowned much of the evidence of ancient coastal civilisations which were destroyed as world ocean level rose by 140 m (most of which occurred about 10,000 years ago). Submarine archaeology of the drowned lands and re-examination of ancient texts (such as the Bible) are now helping us to understand the development of Man during these desperate times.

The modern era has been one of unusual climatic and oceanic stability, allowing stable communities to develop over the last few thousand years – although "mini ice-ages", linked to changes in oceanic current patterns, have certainly occurred. The best-documented of these occurred in the 15th–17th centuries. The onset of this global cooling episode, shortly after 1400 AD, probably wiped out the "Viking" (well, farming, fishing and whaling, anyway!) colony in western Greenland. The Greenland colonies had been established by Erik the Red during the more favourable climate of the 10th–11th centuries; later in this warm period, Erik's son Lief the Lucky colonised north America. By the late 16th and early 17th centuries, the climate of NW Europe was cold enough to allow ice-parties on frozen rivers in England and the Netherlands. The explorer Barents, who "discovered" Spitzbergen and Novaya Zemlya, was trapped on the latter island (and died in unknown circumstances) as a result of local ice-cap expansion during this "mini ice-age". Nowadays, Novaya Zemlya is accessible via the Barents Sea thanks to the present flow direction and temperature of the Gulf Stream (although it is now somewhat unwelcoming, since the Soviet authorities used it as an open-air atomic testing site).

Obviously, the current relative stability of our environment cannot be expected to last.

Such dramatic changes in the environment have tempered Man into perhaps the ultimate non-specialised survivor species. It remains to be seen whether technological humans will survive the end of the present, relatively stable, episode in Earth's history. Perhaps technology can be used to do something about the planet's unstable environmental systems, before it is too late?

"Doing something" would, of course, be extremely difficult and uncertain at our present state of knowledge. That is why it is so important to understand the whole global geoclimatic system and to do that we need more data. The main source of information is the past – and the recent development of tools for detecting the effects of Milankovitch climate cycles on ancient environments will help. That will be the subject of our second snapshot from the past.

Feeling the pulse of climate change – subsurface data from southern Yemen

A study of oil fields in southern Yemen operated by Canadian Petroleum Ltd ("CanOxy") began in March 1998. For ten years or so, the area has been producing copious amounts of oil from a densely-packed group of fields located in harsh terrain, even by Yemeni standards, where oil exploration is tough and expensive. By a stroke of good fortune, the best fields are right where it was easiest to do the seismic data gathering and drilling stages of initial exploration. Oil was struck in the first well, drilled into a reservoir which is now known to extend for tens of kilometres southwards along the main accessible valley but which (for a number of interconnected reasons) seems to be significantly less productive outside this area. For logistical and commercial reasons, the exploration phase of drilling passed very rapidly into the production phase, so a large amount of data was gathered by CanOxy without any immediate need for detailed analysis.

The main objective of my own involvement has been to understand the history of the central group of oil-fields, particularly 19 "key" wells in the Tawila, Camaal, Heijah and Sunah areas (map 3). They were drilled through

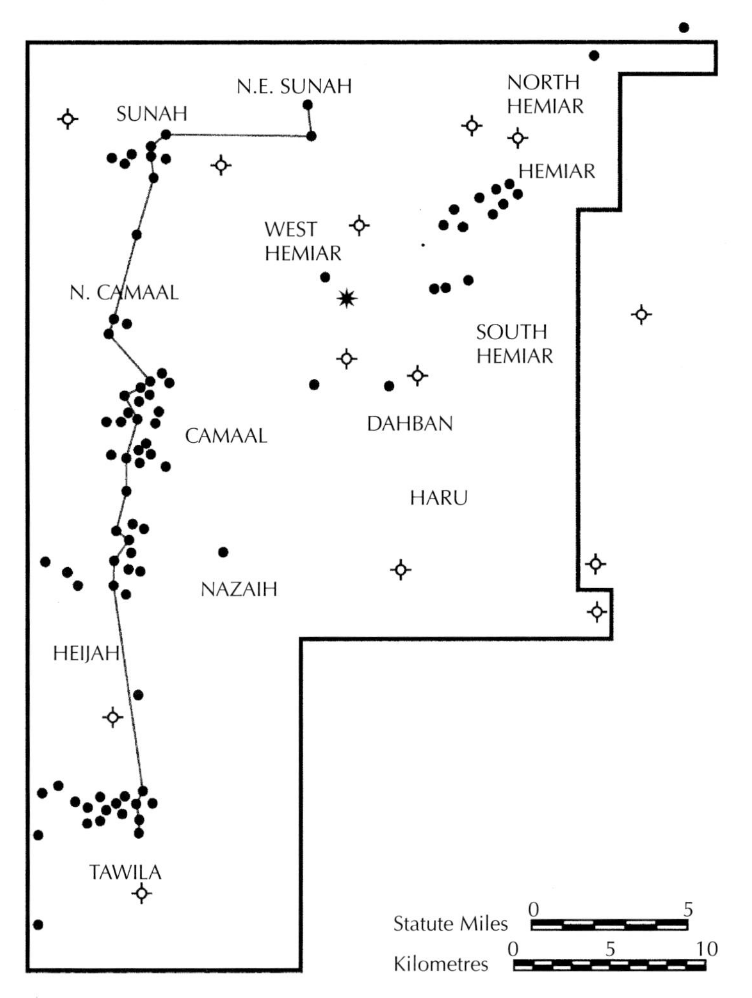

Map 3 Study area in southern Yemen, showing major oil fields, oil wells (black circles) and dry test wells (open symbols). The north–south line connects the wells used in the study described in the text.

the producing reservoir rocks and in some cases cores were cut through part of this sequence. Most of the available data is in well logs gathered soon after drilling. Detailed core descriptions had already been made by CanOxy geologists and I had access to cores stored in Calgary. Access to such a wealth of data is an unusual luxury in my business and permission to publish some of the results even rarer: my thanks are due to Ed Bell and Canadian Petroleum (and their partners) for that.

Southern Yemen is now a desert located on the southern edge of the Arabian Plate to the north of the Gulf of Aden (map 4). The Gulf opened about 15 million years ago as part of the ongoing break-up of a former supercontinent, *Gondwanaland,* which consisted of what is now South America, Africa, Antarctica, Australia, India, Arabia and some other bits and pieces. Sea-floor spreading occurred in the Gulf of Aden along a series of preexisting faults which had a roughly east–west trend, and detached Arabia from the Horn of Africa. These deep-seated faults had formed much earlier near the end of the Jurassic Period (about 150 million years ago) during the creation of a network of subsiding rift-valleys on the north-eastern margins of Gondwanaland. Because of changes in the plate-tectonic motions in the region, this Jurassic partial-rifting stopped before the continent had totally split apart. The inactive ("failed") rift valleys began subsiding more slowly during cooling of the stretched, heated continental crust. This "thermal sag" phase lasted throughout the depositional history of the oil-field reservoirs in the early Cretaceous Period during a time of gradually rising global sea-level, caused by a gradual increase in the speed of plate-tectonic movements and consequent bulge of the mid-ocean spreading ridges (which swell up when they get hot, during rapid plate motion, and displace ocean water onto the continental margins). The early Cretaceous sea-level rises were episodic rather than continuous, and were also driven by other factors; they were generally slower and less dramatic than the large-scale flooding events of the last 15,000 years, and probably had a different origin from those due to the melting of large polar ice-sheets in today's glaciated world.

The rocks of interest to us are now about a mile (1.6 km) below the surface and are known only from subsurface information (diagram 36). These rocks include marine sediments containing oyster fragments, which were dated by strontium isotope analysis at about 122 million years old. The marine sediments were deposited in the late stages of a gradual rise of sea-level, during which a low-lying land surface crossed by rivers was flooded by the sea, which was encroaching from the east. The cores provided enough information, mainly from sedimentary structures and burrows, to prove that the river channels became more tidal with time, eventually being buried under estuarine and shallow marine sediments as the land subsided and was

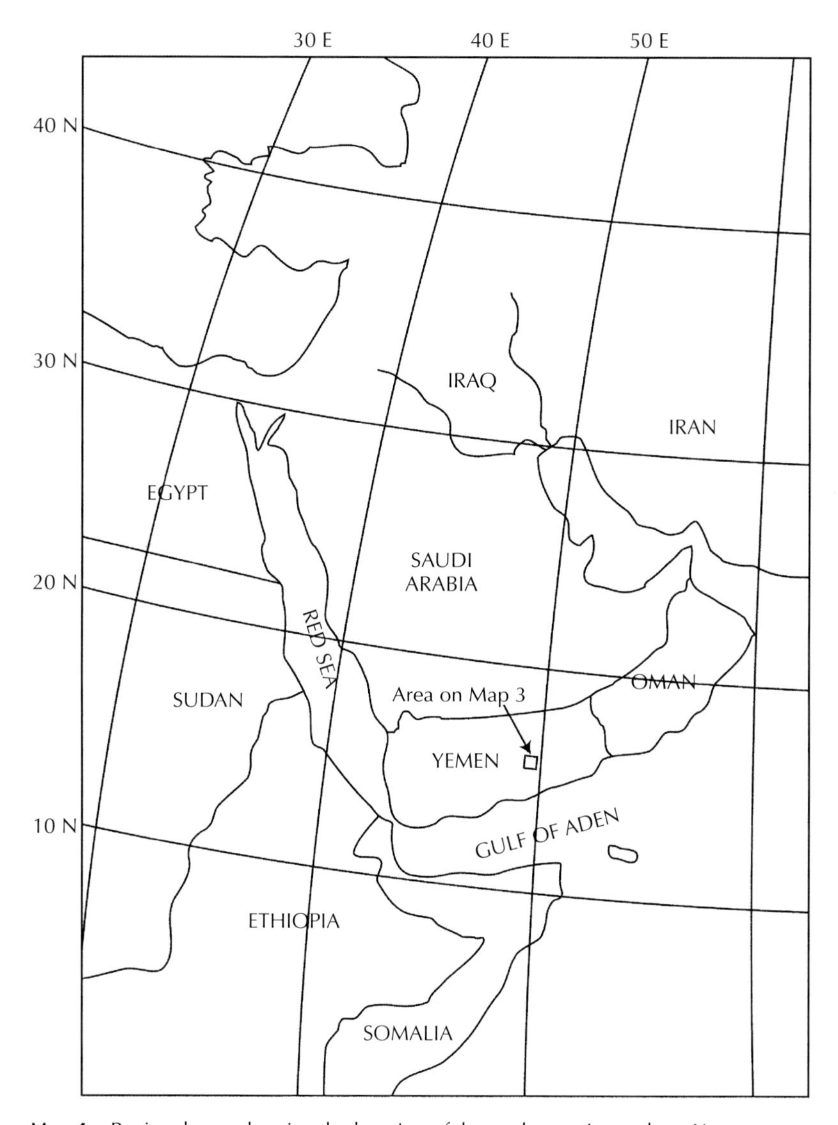

Map 4 Regional map showing the location of the study area in southern Yemen.

drowned by the sea. This change from river to sea took place within a sequence of rocks which is about 50 m thick in all of the wells. This succession is overlain by another 50 m of rocks consisting of shallow marine and coastal sands, muds and limestones deposited in shifting tidal bars and storm beaches (diagram 34).

Oil is produced from clean, porous sandstones which occur throughout this 100 m thickness of rock and is prevented from escaping by a thick capping of dense, non-porous marine limestone. The rocks were folded during the earth movements associated with the opening of the Gulf of Aden, at which time oil migrated into the structures.

The CanOxy geologists had looked closely at the well logs recorded from these wells and matched them with the cores, so it was possible to see which rock types were responsible for different sets of log signals (diagram 34). The logs from adjacent wells were very similar in most respects, showing that the layers of rock were fairly continuous. Thanks to a consistent data-set

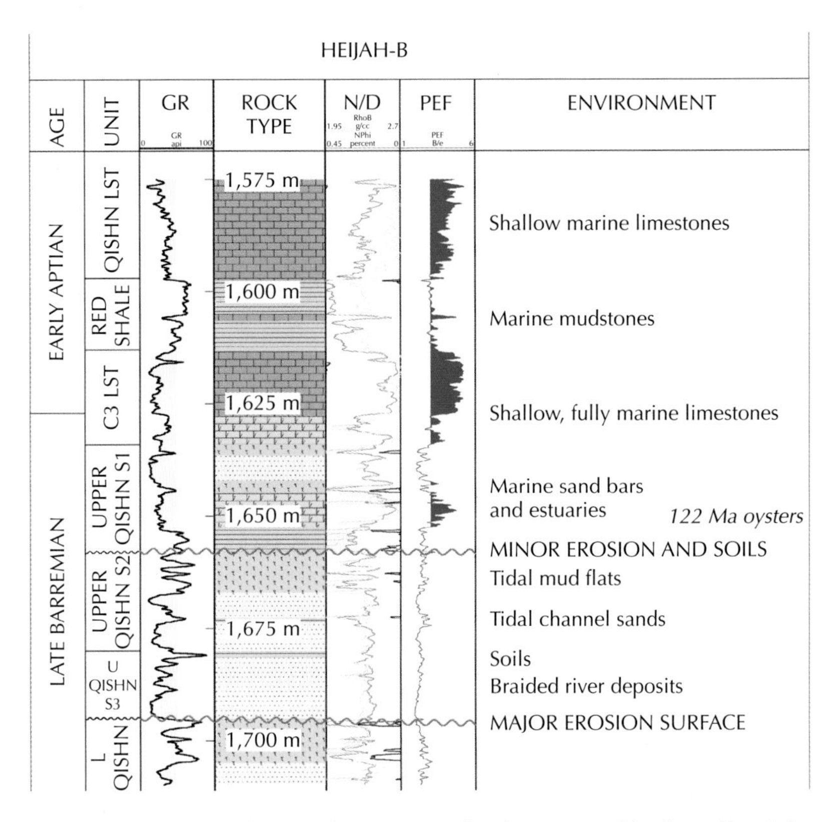

Diagram 34 Diagram showing the sequence of rocks penetrated by the well Heijah-B, 1575 m to 1700 m below the surface of southern Yemen. The well logs (GR = gamma ray, N = neutron, D = density, PEF = photoelectric log) were used to define the rock types in the well (see Chapter 3). Details of the environments of deposition were worked out from examining cores and cuttings. The names of the rock units and their ages are mainly based on comparisons with wells and outcrops elsewhere.

and good computer software, it was easy to define a clear log correlation between the wells, based on the main log trends and patterns (diagrams 35 and 36). The pile of rock resembles a "layer cake", with each layer (defined by the lines connecting the wells on the correlation diagrams) being no more than a few tens of metres thick. The close similarity of the log data within each layer in every well indicates that these layers record fairly uniform conditions from north to south, along 30 km of the ancient coastline of this slowly subsiding, flat-lying region. It is reasonable to suppose that each layer was deposited at much the same time in each well.

It soon became obvious that this data set offered a perfect opportunity to test the CycloLog computer program, which uses Maximum Entropy Spectral Analysis to define cyclicity in well logs (see Chapter 5), and to see whether the well logs contained a regular pattern of signals generated by periodic climate changes (Milankovitch cycles). If it were possible to identify the particular Milankovitch cycles responsible (such as the 100,000-year eccentricity or the 40,000-year obliquity cycle), we would know exactly how old various layers in the wells were, compared with the 122 million-year old oysters, and how rapidly the sediment pile had accumulated. Such parameters had never before been quantified with this precision.

Right at the beginning of this book, I commented that it is never possible to prove anything absolutely in science. Instead, one has to disprove all alternative realistic hypotheses, leaving just one reasonable explanation as the surviving theory. The critical part is deciding what is reasonable or realistic and what is not. Thus, we could not expect to *prove* that ". . .the well logs contained a pattern of signals generated by periodic climatic changes" but we might be able to show that this was the only reasonable explanation for our observations. A test needed to be designed to show that reasonable conclusions could be derived *only* if cyclic patterns detected in the 19 wells were due to regular Milankovitch cycles.

The CycloLog program was used to find out whether repeated, "cyclic" patterns exist within the log signals in each well. The analysis of each well showed that this was the case, and displayed the average thickness of the cyclic, repeated patterns for each set of log signals (diagram 37). These cyclic patterns are bounded by "peaks" or "kicks" in the log data, which are a response to unusual rock types. In the gamma-ray logs, these peaks occur in muddy shales which are rich in radioactive material. The cycle wavelength is the thickness between these "hot" shales. In the well shown in the diagram, there are 12 such peaks, with an average spacing of about 7.6 m. They show up in the gamma-ray data and also in the analysis of Neutron/Density data, which is an alternative way of identifying "hot" shales, providing a check on the accuracy of the gamma-ray analysis.

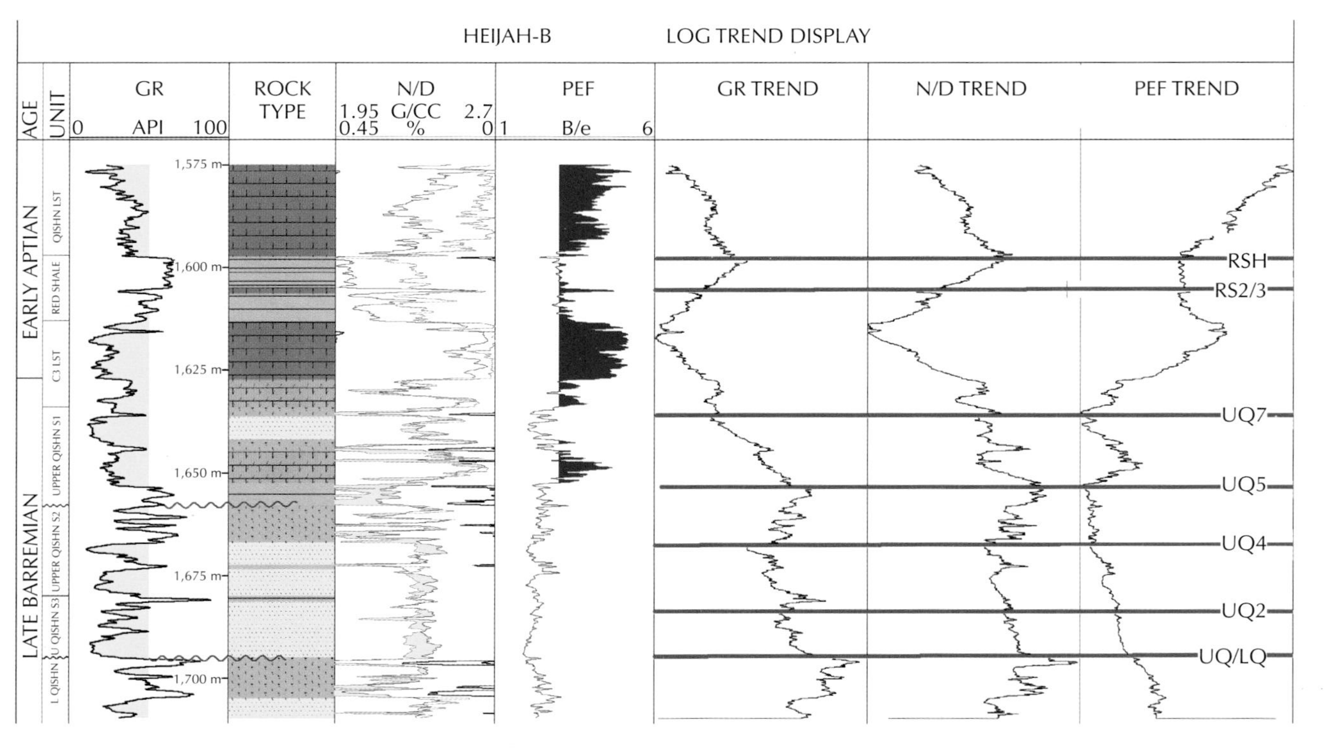

Diagram 35 Gamma ray, neutron, density and PEF logs from the Heijah-B well, together with plots of the main trends in these logs derived from an analysis using the CycloLog computer program. The shapes of the trend plots are used to define marker levels shown in red. These markers can be recognised in the other wells in the area.

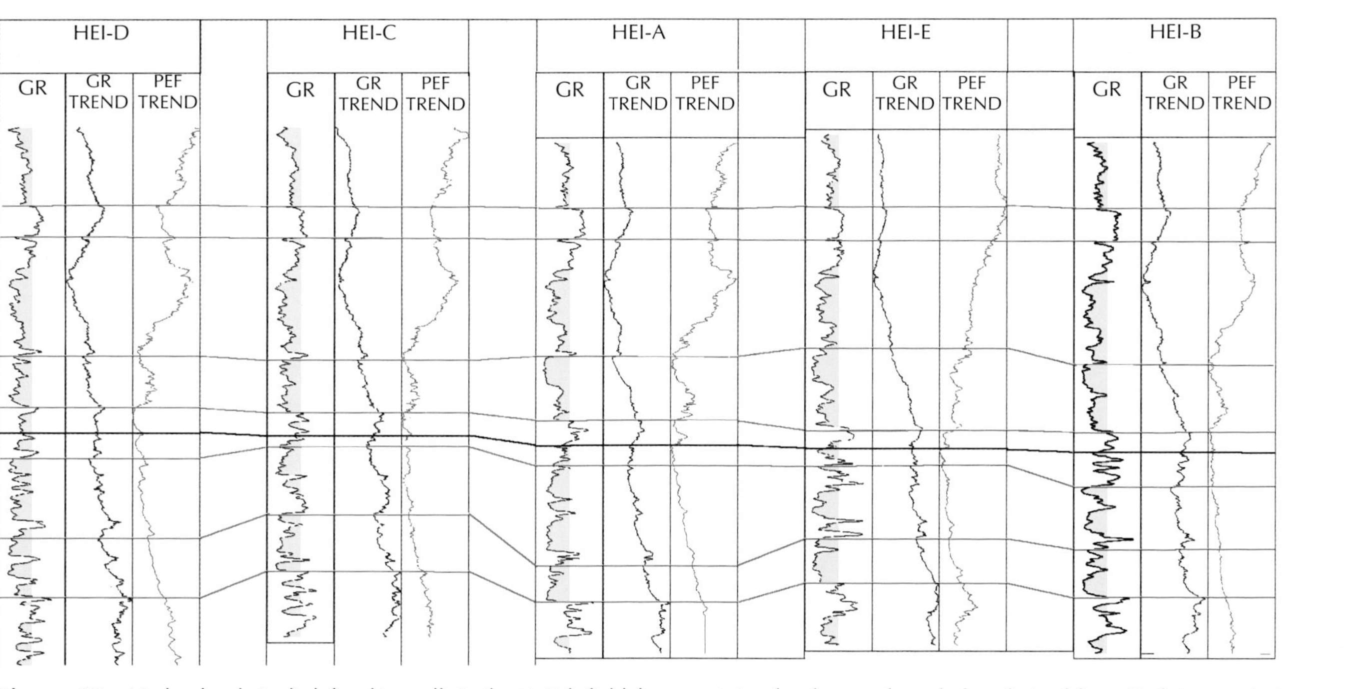

Diagram 36 Marker levels (red) defined in wells in the Heijah field, by examining the shapes of trend plots derived from CycloLog analysis of the gamma ray and PEF logs. The black line is an erosion surface with soil development, which is seen in cores from many of these wells.

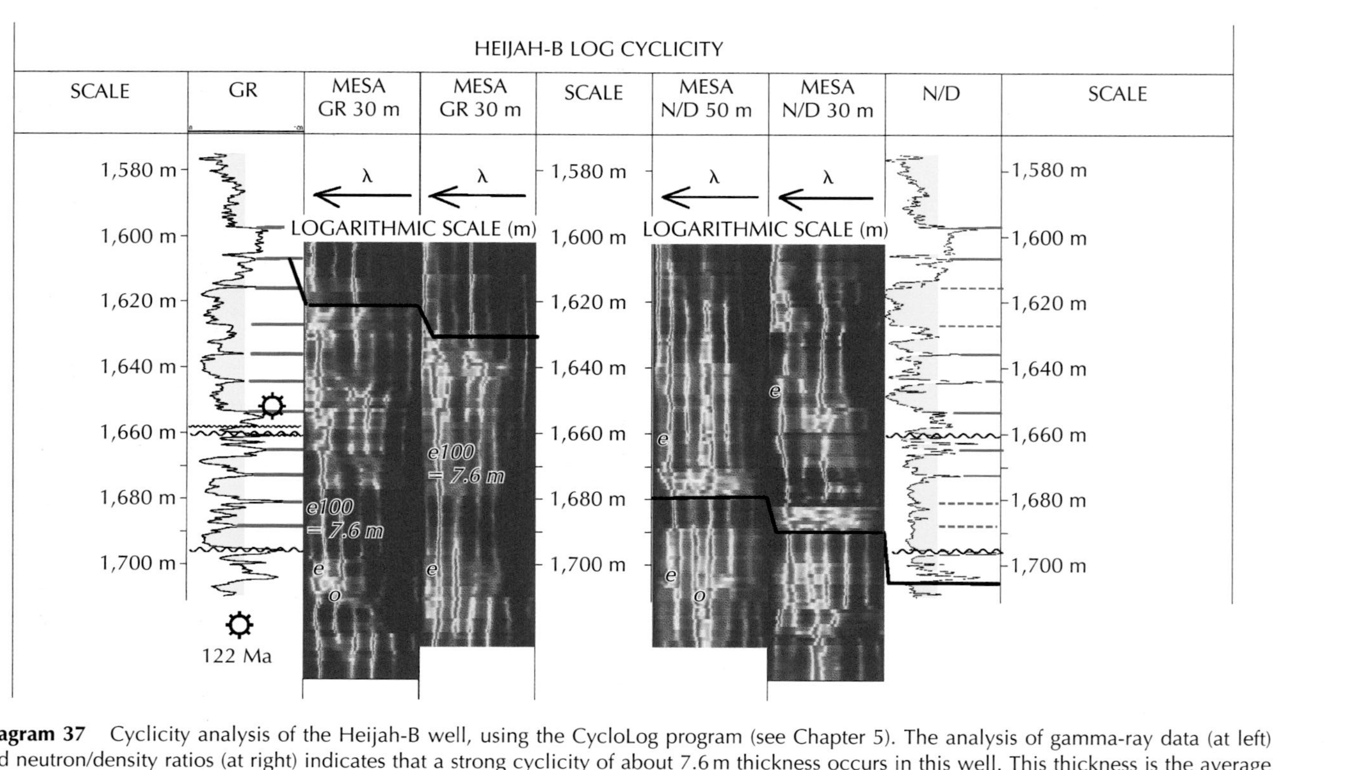

Diagram 37 Cyclicity analysis of the Heijah-B well, using the CycloLog program (see Chapter 5). The analysis of gamma-ray data (at left) and neutron/density ratios (at right) indicates that a strong cyclicity of about 7.6 m thickness occurs in this well. This thickness is the average separation of "kicks" in the gamma-ray and neutron/density data, indicated by red lines. The "kicks" are seen in core to correspond to organic-rich muds, which were deposited during rapid marine flooding of the area at times of rising sea-level.

Core samples showed that these hot shales contain evidence of rapid flooding of shallow seas over the widespread tidal flats on which most of the other sediments in the wells were deposited. The evidence comes mainly from the traces of burrowing organisms, which flourished in the flooded, shallow marine areas.

So, the cyclicity analysis was detecting the spacing of shales deposited by marine flooding events. In most of the wells, the same number of flooding events was recorded within each correlated part of the overall "layer-cake" sequence (diagrams 38 and 39). However, the result was not quite as straightforward as I'd hoped. In parts of the rock succession in some wells, the flooding events were difficult or impossible to detect from the cyclicity analysis – these parts are marked on the diagrams. The thick cycles indicated in these sections were not imaged by the analysis, and I had to project them into these wells by comparing the log patterns with those in nearby wells. However, these sections did contain thinner cycles – generally about two fifths of the thickness of the thick cycles shown on the diagrams. We will return to this shortly.

The result suggested that the whole area, measuring some 30 km from north to south, had been flooded by the sea fairly rapidly on a number of occasions during the deposition of each layer in the "cake". The fairly regular thickness between the gamma-ray peaks in each well suggested that these flooding events had occurred at approximately constant intervals in time (assuming relatively constant subsidence in each area, which seemed reasonable based on our knowledge of the long-term history of the region). Minor variations in the thickness of sediment between the flooding events from well to well could be explained as changes in the rate of subsidence across the region.

So, if these flooding events were fairly regular events, what could have been causing them? Although there are numerous possible explanations, I wanted to test the hypothesis that they were controlled by the regular climate changes caused by Milankovitch cycles.

A total of 15 of these regular events are present in most of the wells, and the same number is present in each of the widespread layers in the "cake". If we assume that the events were controlled by a particular Milankovitch cycle (the 100,000-year eccentricity cycle, for example), we can count off the number of such cycles in each layer to work out the length of time it took to deposit each layer.

Unfortunately, this is inconclusive. Although the isotope dating of the oysters near the top of the sequence placed that layer at about 122 million years old, there is no good evidence of the age of any deeper part of the sequence. Even if there were any dated fossils lower in the wells (which there

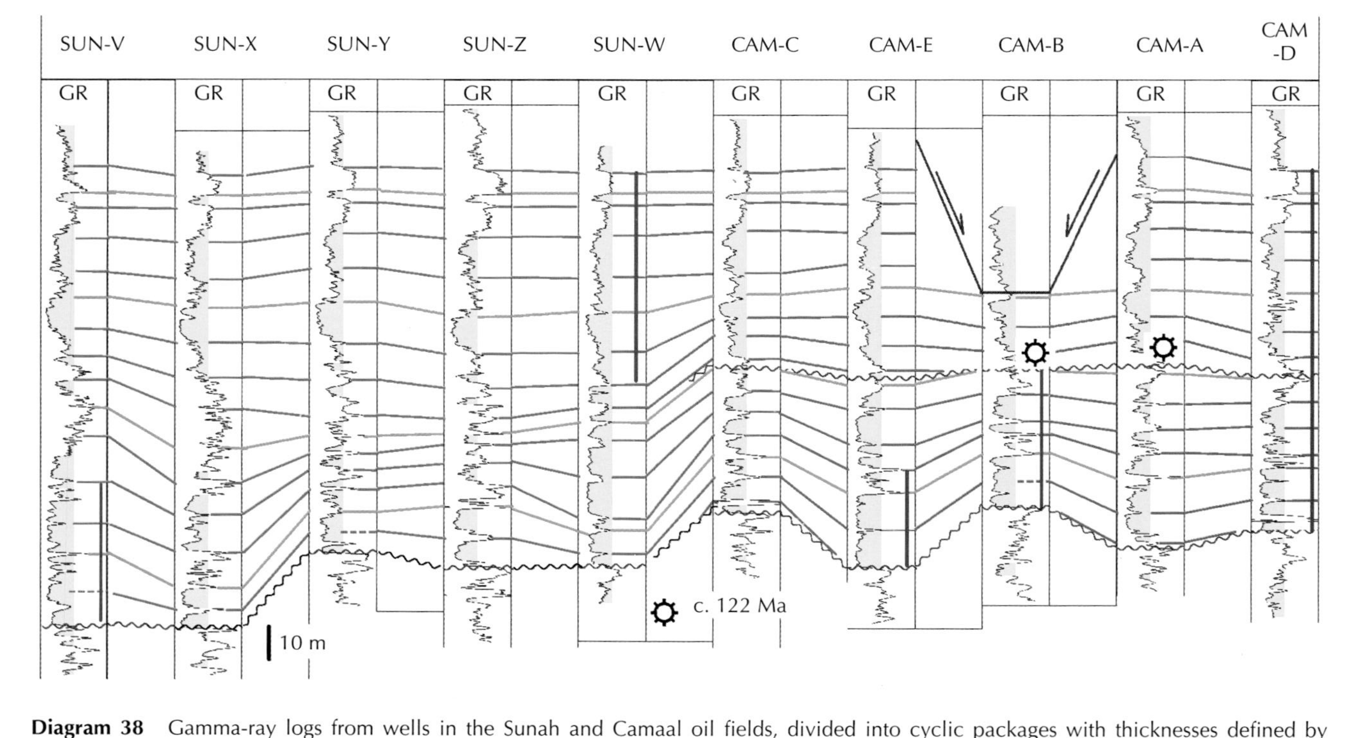

Diagram 38 Gamma-ray logs from wells in the Sunah and Camaal oil fields, divided into cyclic packages with thicknesses defined by CycloLog analysis. Every fourth marker is coloured green (see text for discussion). The wavy lines mark erosion surfaces where some of the sediment succession is missing. A large section of the Camaal-B well is cut out by a fault. The vertical blue lines show where the cyclic packages were not imaged by CycloLog analysis, although thinner packaging was recorded (see text).

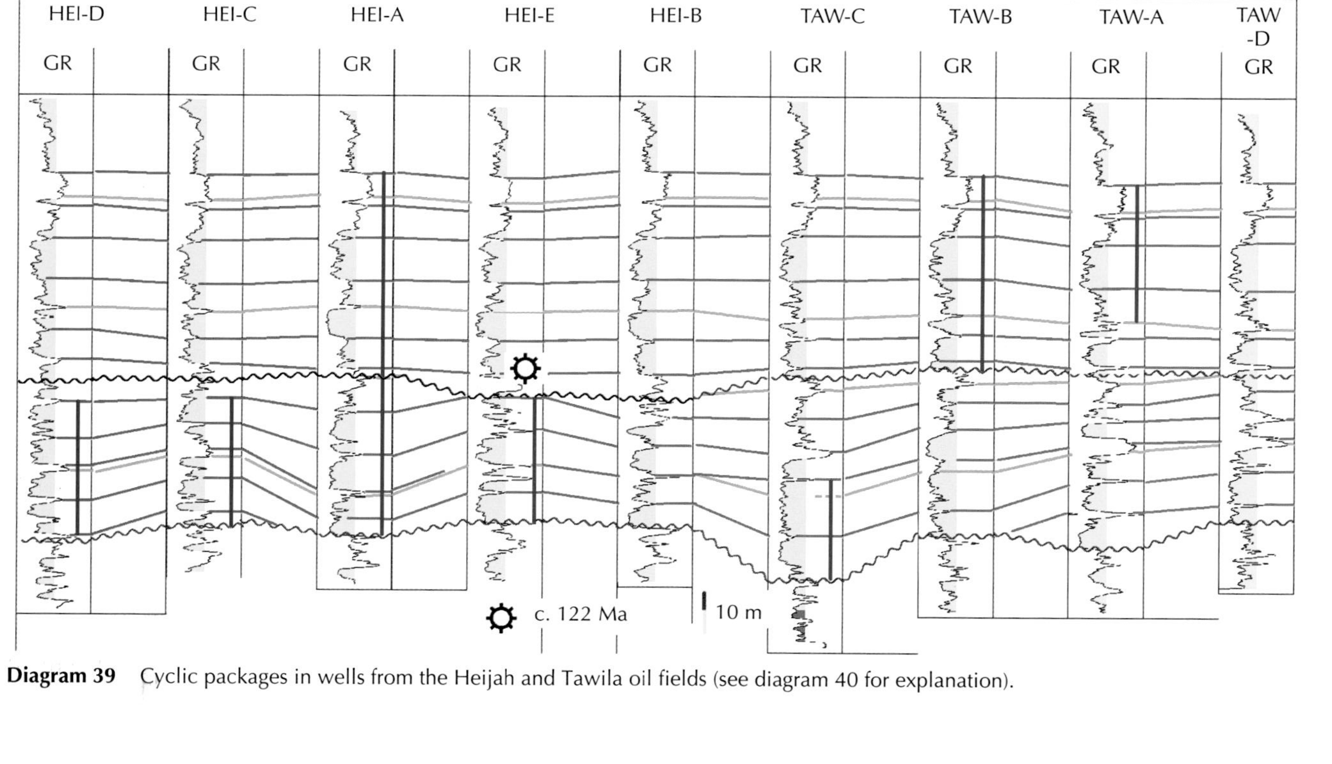

Diagram 39 Cyclic packages in wells from the Heijah and Tawila oil fields (see diagram 40 for explanation).

aren't), fossil dating is not precise enough to decide whether the wells actually record 100,000 year flooding cycles, 40,000 year cycles, or something else.

This is where the other, thinner cycles detected in many of the wells become important. These were often not as strongly detected by the spectral analysis as the main flooding surface cycles, but in some wells they are the only cycles detected by the computer program. Their thickness is typically two-fifths of that of the main, thicker cycles (diagrams 38 and 39). In the wells containing only thin cycles within part of the layer cake, I was able calculate the thickness of "undetected" thick cycles by multiplying the thickness of thin cycles by 2.5 (diagram 40). The match between the well logs produced in this way is so good that it seems obvious that the thin cycles are real, and that they always have this 2.5:1 thickness ratio with the main flooding cycles. If we assume that the cycles are of Milankovitch origin, then the thick cycles probably represent 100,000-year eccentricity-driven climate changes and the thin cycles probably represent 40,000-year obliquity-driven climate changes, because these two Milankovitch cyles are the only ones with a 2.5:1 ratio.

Based on this argument, we can draw lines across the region, joining up the 100,000-year gamma-ray peaks to trace the record of major marine flooding events from well to well (diagrams 38 and 39). The layers truncate at an erosion surface within the layer-cake sequence, when the southern part of the area was exposed above high-tide mark and eroded, usually with ancient soils preserved at this level in the cores.

The final piece of evidence which suggests that these Milankovitch cycles are real, comes from the pattern of 100,000-year cycles in the wells. We might expect every fourth cycle to be more important, due to the 400,000-year "beat" in the eccentricity cycles. There is a thin limestone in all of the wells, near the top of the sequence. This is present just below 1600 m in the Heijah-B well (diagram 36) and can be traced throughout the region, for hundreds of kilometres. This marine limestone layer represents a maximum flooding surface (see Chapter 5), which records the most widespread of all of the flooding events. It is highlighted on the correlation panels, capping an unusually thin cycle (which is probably thin because deposition was slow at this time of high sea-level, due to reduced mud supply from the land). Four cycles below that is a major flooding event (also highlighted), which terminated deposition of the sandstones which form the top of the oil-bearing reservoir (about 1630 m depth in diagram 34). Another four cycles down is an unusually "hot" sequence of shales, which probably represent a time of extensive marine flooding (best displayed in the Sunah-V well at the left of diagram 38). Four cycles below that is a change from

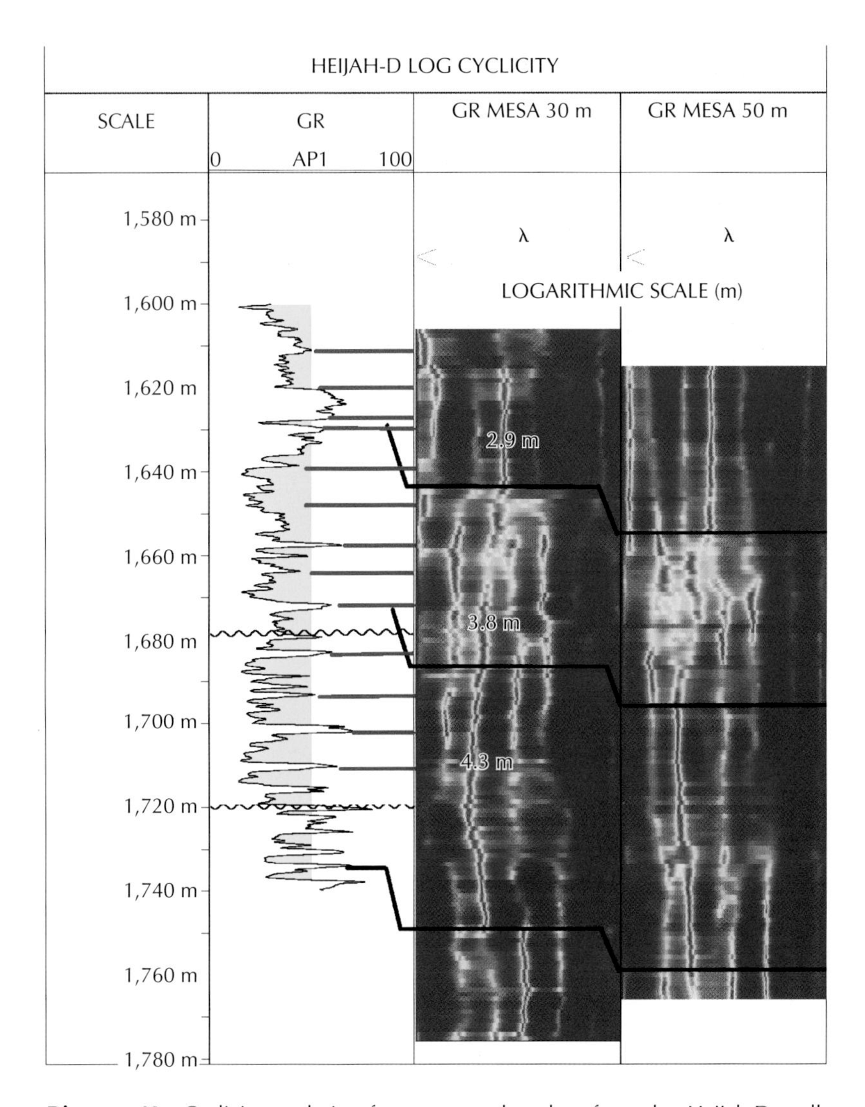

Diagram 40 Cyclicity analysis of gamma-ray log data from the Heijah-D well, showing thin packages (4.3 m, 3.8 m and 2.9 m). These are about 2.5 times thinner than the packages marked by the red lines, which correspond to the thick cyclic packages seen in other wells (compare with diagrams 38 and 39).

low-gamma sands to higher-gamma muds at the top of the basal sequence of river deposits, marking the first major sea-level rise when the river system was inundated by the sea to form tidal creeks and estuaries. These major changes seem to have occurred at 400,000 year intervals, according to the present interpretation, which is consistent with a Milankovitch origin for the log cyclicity.

So, how could Milankovitch climate cycles have influenced the deposition of this sequence in such a profound way? The simple answer is that we don't really know. The cores from these wells could be studied in future to investigate other evidence of climate change in the sediments – from assemblages of fossil plant spores and pollen, or from changes in the clay minerals weathering from the land, for example.

At present, we can speculate that changes in global temperature driven by the Milankovitch cycles 122 million years ago caused the oceans to expand and contract in volume as they heated and cooled, periodically flooding the flat coastal landscape as sea-level rose and fell by a small amount (probably a few metres at most). There is some evidence for polar ice caps at that time, which would provide a way of locking up sea-water as ice during cooler climatic phases. Or sea-level could have been controlled by other factors, such as the amount of rain falling in the continental interiors to be stored as groundwater and lakes (which today constitute less than one percent of the water on Earth – about 97% is in the sea, and 3% in the ice caps). Alternatively, global sea-level may have remained fairly constant, and the subtle changes in the landscape in Yemen at that time may have been driven by the amount of sediment being washed off the Arabian interior by rainfall, which varied according to the prevailing climate.

The main point is not that we can understand the development of southern Yemen 122 million years ago. The study has wider implications. It demonstrates that Milankovitch cycles can be detected in well log records from rocks deposited before the start of the Ice Age (i.e. older than about 30 million years old). This kind of study is now being performed elsewhere, on other ancient sequences penetrated by oil and gas wells, with equal success.

The precise definition of 100,000-year units in the rocks in these wells is beginning to build up an incredibly detailed picture of the development of ancient landscapes over time, showing us how the environment in each area responded to the regular climate changes forced by Milankovitch cycles. We do not fully understand the environments or the climatic effects at present, but we now have the opportunity to improve our understanding, guided by very precise age dating of ancient examples.

7

The big picture

Several excellent popular Earth science books published recently outline the main events in Earth history and highlight the major catastrophes which have helped to shape our ecosystem. Their general message needs emphasising: although gradual "everyday" processes are the norm, we need to be more aware of the unusual or "catastrophic" events. There are several reasons for this.

First, we base much of our scientific understanding of the planet's working on observations of events that occur during our own lifetimes in reasonably accessible areas. We therefore know a lot about how slow, lowland rivers operate and a lot less about the dynamics of upland torrents fed by bursting glacial dams. Deep-sea sediment transport is equally difficult to study at first hand. However, given a defined set of parameters (slope, water viscosity, sediment concentration and so on) and scaled-down models in the laboratory, we can make some sensible predictions. But, unless we find evidence of past events which we cannot explain by "everyday" mechanisms, we rarely appreciate what the Earth is capable of in extreme circumstances. That can make it difficult to recognise unusual events in the geological record because their effects tend to be (wrongly) interpreted in terms of more common processes. This is one drawback of what scientists call "Occam's Razor" – the principle that the simplest, least "unusual", explanation is generally the best. For example, we are almost certain that occasional impacts of large asteroids or comets into the oceans have generated giant tidal waves (*tsunamis*) which swept across the lowland areas of nearby continents. On theoretical grounds, this is thought to have occurred every million years or so. And yet we know of hardly any convincing examples from the geological record. Even the supposed record of a tsunami which swept across southern Texas at the end of the Cretaceous, caused by the impact of the comet or asteroid which finished off the dinosaurs, is disputed by many sedimentologists – and in that case, we really do know that a large object hit the sea not far away. It is even more difficult for most scientists to accept that other deposits of jumbled, chaotic shallow-marine sediments were caused by impact-related tsunamis, when it is so much safer to ascribe them to large storms or earthquakes. So how can we study the effects of such unusual events, and perhaps plan to deal with them in the future, without convincing examples to work on?

Second, because the record of thousands of years of slow deposition can be swept away in a single major storm, landslide or other energetic flow, it is a generally accepted fact of Earth science that most of the geological record is dominated by unusually powerful events. Long periods of "nothing much" happening can be represented by a thin mud layer, whereas a thick sand layer may be deposited in a few minutes. In fact, sedimentation has been likened to war: long periods of boredom are punctuated by brief moments of terror. So it is particularly important for the sedimentologist to be familiar with "unusual" sedimentary processes, in order to understand a significant proportion of sedimentary rocks. This fact seems to have escaped a few geologists who prefer a quiet, ordered picture of slow accumulation of sediments, in conditions you might see on a fair-weather field trip.

Third, and more importantly for most of us, it has become clear that Man has evolved through adaptation to a series of drastic, catastrophic events which have forced life on Earth to adapt or perish. It is hardly fair on the dinosaurs to imagine that they died out simply because they could not adapt very well – although individual species lasted for much shorter periods, the dinosaurs were the dominant animals in their ecosystems for 150 million years, surviving some major catastrophes along the way. But the disastrous end of the Cretaceous was too much for them. Global sea-level fell, reducing the area of shallow shelf seas and killing off many fragile ecosystems; global climate became less hospitable; vast volcanic eruptions in India poisoned the atmosphere; and just when things seemed they could not get any worse, an asteroid or comet hit Mexico and blackened the sky for a year or more, setting off continent-wide wild-fires and filling the air with sulphurous fumes. The surviving animals seem to have been those which could burrow, hibernate, scavenge and operate well in the dark. Unless you include their relatives the birds, the dinosaur species which lived at that time could not do what was necessary to survive in this harsh new environment. But eventually the "good times" returned; mammals became dominant and giant rhinos and other exotic beasts evolved their own "dinosaur"-like specialisations.

The specialists always did better when life was fairly easy, but come the next comet or other catastrophe and it was usually the generalists which survived. They were ready to expand into the vacant ecosystems once the killing was over and some then evolved their own exotic specialities until the next disaster struck. Are there lessons in this for us?

Man has evolved during a series of catastrophes, namely the Ice Ages and the related effects of plate tectonic collisions, the uplift of the Himalayan/Alpine uplands, the eruption of supervolcanoes in Indonesia and elsewhere, and dramatic changes in global oceanic circulation. Being both adaptable and intelligent, humans, perhaps the most opportunistic of mammals, have survived and spread throughout the world. Modern Man has now lived through an unusually long period of "good times" of relatively stable conditions, and has developed advanced communication and technology. It could be argued that this is a "dinosaur"-like specialisation, which cannot be expected to survive any major catastrophe which disables our current life-style.

Perhaps there is a way to deal with the statistically increasing threat of catastrophic change – risky and difficult, but still possible: a solution based on science and aimed at continuing the present development of our society. We can try harder to understand the workings of the Earth, to predict coming ecological disasters and then take action to control or minimise their effects, by intervening in some way. The term "terra-forming" was coined to describe how one might go about turning Mars or some other planet into a habitable, Earth-like place, using microbes and chemical catalysts to alter the atmosphere, solar reflectors to increase the amount of sunlight and so on. Will we need terra-forming to be able to continue here, never mind about on Mars? The problem will be to judge precisely how any action we may decide to take will affect the myriad unstable systems it acts upon.

Most terrestrial systems are "chaotic": they tend to change slightly all the time but stay close to some "stable" condition until they are jolted out of that small range of variation into a radically different stable condition. In chaos theory, these two or more stable conditions are called *strange attractors*. Think of a ball-bearing sitting in one of a series of shallow dimples on a flat board. Nudge the board a little and the ball-bearing will wobble in its hollow but drop back into place. Nudge it too much and the ball rolls off until it eventually settles into a new hollow somewhere else. Our weather is like that. It varies a lot in a rather unpredictable manner but stays close to some norm. We know a lot about the current norm (it generally rains in Britain at some time in December) but it is impossible to predict exactly how the system will deviate around the norm in any detail (in November we have no idea whether it will rain on Christmas Day). This weather pattern is one stable condition of a chaotic system. The general climate of Britain has been stable for the whole of my life but if the Gulf Stream stops bringing 25% of the UK's annual heat budget north-eastwards from the Caribbean, I can confidently predict that we will not have rain in December – because then it will be too cold for rain and we will have snow, lots of it! And that could happen in the lifetime of anyone reading this book, if the current melting of the Greenland ice-sheet cools and deflects the Gulf Stream enough.

Ironically, even the "regular" Milankovitch cycles themselves are a classic example of a chaotic system. On a "short" timescale (the 30 million year duration of the present Ice Age, for instance), the whole system is almost perfectly cyclic and predictable, but it cannot be expected to stay that way. An exceptional alignment of the planets, or the influence of a passing star, could cause minor but significant changes in the currently stable orbit of the Earth and the variations of its axial tilt. Such changes would alter the duration and magnitude of the Milankovitch cycles. Fortunately for our purposes, this does not appear to have happened over the last few hundred million years.

The Earth's ecosystems and climate are an incredibly complex chaotic system, like a board with hundreds of ball-bearings and thousands of dimples. Jolt the board too much and the whole system will rearrange in a highly unpredictable manner. We need to know what to expect from any given tilt of the board but we have hardly begun to find out.

Understanding the variations in our climate is hampered by a lack of clear examples of what has happened in the past. This is because most of the geological record is too poorly dated to give a precise picture of cause and effect, to show how a particular change in conditions has worked through the myriad interacting systems of the Earth to produce some end result. Computer models of global climate are still far too primitive to allow purely hypothetical modelling of anything but the simplest of changes in the system. Understanding the record of past changes is the key to predicting changes in the future in this respect. What we need is an improvement in dating the geological record and in understanding the effects of ecological changes that have occurred in the past.

An important breakthrough may now have been made. By using the record of Milankovitch climatic cyclicity, revealed by computer analysis of geophysical data, we can begin to date sedimentary sequences with an accuracy previously unattainable. The effects of Milankovitch climate cycles can be traced across the non-glaciated world and the Ice Age world to see how the predictable, regular changes in incoming sunlight are filtered through the chaotic systems of the environment to produce measurable changes in the geological record. By comparing well-dated records from the same timeframe in different parts of the world, we can see how local geography or oceanography have altered the local effect of global climate change. The Milankovitch cycles detected in well logs can also provide accurate relative age dates which should help us to understand the response of Earth systems to other events. From such detailed, painstaking and expensive research, we should be able to develop a far better understanding of how the Earth will react to future change and how the myriad ecosystems might respond to any attempts to control those changes. Expensive and difficult work – but luckily this information is a by-product of the global energy industry, and it is now being acquired. But would it be too risky to try to control climate change and other ecological disasters?

Well, what happens if we do nothing, other than to protest vaguely about the inevitable consequences of civilised existence? We know a fair amount about global warming, the greenhouse effect and depletion of the ozone layer, but very little about the catastrophic global changes worked by supervolcanoes (one of which nearly wiped out mankind 76,000 years ago), magnetic reversals, changes in ocean circulation (such as the millennial-scale oscillations of the Atlantic Gulf Stream in response to changes in melt-water flow from North America) and the occurrence of tsunamis (generated by earthquakes, comet/asteroid impacts or the large-scale collapse of submarine volcanoes and other unstable slopes). It would be foolhardy to try to remedy one problem if our actions exacerbated another, linked effect. A little knowledge is a dangerous thing. But ignorance is worse.

For example, few people know what causes the so-called Bermuda Triangle effect, where aeroplanes and ships have suddenly disappeared without trace since records began, in a large area of coastal waters off the southeastern USA. This is now thought to be due (at least in part) to the sudden, catastrophic release of large volumes of natural gas (methane) from the deep sea-floor. Methane seeping from deep underground mixes with sea-water under high-pressure, low-temperature conditions (as in the deep sea) to form a solid clathrate known as *methane hydrate*. If something changes the stability of the hydrate (perhaps an increase in temperature, a decrease in confining pressure, an earthquake, or the activity of a sea-floor dredger such as a whale) then some of the gas is released. Similar events in the Caspian Sea have been known to science for decades. In the Bermuda Triangle, large areas of the sea-floor are shallow enough for any hydrate layers to be close to the depth at which pressures are low enough to release the methane contained in them. A large earthquake could result in the sudden release of large volumes of methane, locally turning the sea into a gas-charged foam. Ships cannot float in this foam, and planes flying overhead can only fly in air, not in the less-dense gas methane. The stricken vessels sink rapidly to the sea-floor, where they are soon buried as the gas-release ends and disturbed sea-floor sediment settles out of suspension.

Oil companies have known about this hazard for some time, and they routinely survey potential drilling sites to detect shallowly-buried gas hydrate layers before starting deep-sea drilling operations. Of course, the methane bound up in extensive deep-sea hydrate layers could be exploited, but it is generally not concentrated enough to be worth the risk and expense of extracting it. However, it has been suggested that gas produced from deep-sea drilling could be vented at the sea-floor to allow solid hydrates to form, instead of lifting the gas to the surface in pipes. The solid methane–water mixtures might then be transported to more convenient locations for further processing – possibly even by shaping the extruded solids into steerable "submarines"; one could even imagine allowing some of the solid hydrate at one end of such a vessel to release some of its gas, providing a primitive form of propulsion. . . . Such ideas are interesting, but one must remember that methane is a more effective greenhouse gas than carbon dioxide, so releasing more of it into the oceans (and atmosphere) is not environmentally wise. Also, we now know that vast areas of the North Atlantic continental slope have collapsed during the Ice Age, apparently due to sub-sea landslides. These are most probably related to widespread methane hydrate layers in these areas, which de-gassed as confining pressures dropped or sea-bottom temperatures rose, in response to rapid changes in sea-level and temperature as the polar ice-caps grew and retreated.

It is obvious, then, that there are myriad inter-related phenomena at work on the Earth. These need to be considered before any attempt is made to intervene in some perceived problem (such as global warming). Considered and united decisions will have to be taken if we are to keep our house in order, and these will have to be based on much better knowledge of the past, our only reliable key to the future, than we have now.

Suggestions for further reading

Geology and Earth history

Earth Story by Simon Lamb and David Sington, BBC Books, 1998, 240pp.
Lavishly-illustrated spin-off from the BBC series (available on video) which describes the development of ideas about plate tectonics, the origin of the Moon, and the linkage between Earth history and the evolution of life on Earth. Highly recommended – probably the best popular Earth Science book published last century . . .

Sedimentology: Process and Product by M.R. Leeder, George Allen & Unwin, 1982, 344pp.
An occasionally entertaining, if technical, description of depositional processes and the sedimentary deposits which they produce. Excellent starting point for a Geology student – but probably superceded by newer texts since my time.

Sandstone Depositional Environments edited by P.A. Scholle and D. Spearing, American Association of Petroleum Geologists, Memoir 31, 1982, 410pp.
A book packed with excellent photos of what sedimentary rocks look like at the best outcrops in the world, with a clear and readable explanation of what the exposures tell us. Probably a bit too technical for many people, but stunning stuff.

Sequence Stratigraphy edited by D. Emery and K.J. Myers, Blackwell Science, 1996, 297pp.
The clearest explanation (that I know of) of this rather confusing and jargon-cluttered aspect of geology, which has caused almost as many problems as it has solved. . . . This book is basically the published version of a BP training course, and is therefore very well illustrated with real case examples and very clearly written. Essential reading for anyone faced with the task of understanding sequence stratigraphy and with the mountain of literature written on the subject in the last 25 years.

A Dynamic Stratigraphy of the British Isles by R. Anderton, P.H. Bridges, M.R. Leeder and B.W. Sellwood, George Allen and Unwin, 1987, 301pp.
A good example of the use of geological observations at a range of scales to describe the changing environments of a region, in this case one of the most thoroughly studied regions in the world, with an almost continuous record of events throughout the evolution of life on Earth.

Petroleum Geology of the North Sea – basic concepts and recent advances edited by K. Glennie, Blackwell Science, 1998 (4th edition), 636pp.
Highly technical but very well-written description of the development of the North Sea rift system since the Devonian (over 400 million years ago), with fascinating details of the formation of its abundant hydrocarbon deposits. Written by the oil-industry's top experts on this subject, and therefore clearly explained and (often) easy to follow.

The geological interpretation of well logs by M. Rider, Whittles Publishing, 1996 (2nd edition), 280pp.
Clearly-written and definitive explanation of the meaning and use of wireline logs in evaluating the rock types and geological record of oil and gas wells. Essential reading for anyone serious about using well logs. Even mentions spectral analysis of log data in the last chapter!

Milankovitch cycles and climate change

Imbrie, J., Boyle, E.A., Clemens, S.C., Duffy, A., Howard, W.R., Kukla, G., Kutzbach, J., Martinson, D.G., McIntyre, A., Mix, A.C., Molfino, B., Morley, J.J., Peterson, L.C., Pisias, N.G., Prell, W.L., Raymo, M.E., Shackleton, N.J. and Toggweiler, J.R., 1992. On the structure and origin of major glaciation cycles. 1. Linear responses to Milankovitch forcing. *Paleoceanography*, **7**, 701–738.
A highly detailed and technical discussion of the influence of Milankovitch cyclicity on glaciation cycles, with palaeoclimatic models and mathematical analysis of a wide range of climatic data for the last 400,000 years – contains all you could ever want to know, probably.

Matthews, M.D. and Perlmutter, M.A., 1994. Global cyclostratigraphy: an application to the Eocene Green River Basin. In: De Boer, P.L. & Smith, D.G. (Eds.), *Orbital forcing and cyclic sequences*. Spec. Publ. IAS, **19**, 459–481.
A theoretical discussion by Texaco scientists of the possible influence of Milankovitch cyclicity on global climatic belts, causing them to shift latitudinally with time. This helps to explain sedimentary response to Milankovitch cycles in terms of fluvial discharge and erosion rates, in addition to

glacioeustatic mechanisms. This approach is now being used by Shell and others to predict sedimentary rock distributions in undrilled areas. This is a kind of "forward modelling", i.e. you find out as much as possible about what processes have happened in the past, then try to predict what products should have formed. If your prediction resembles what you observe (in this case, the deposits encountered in an exploration well), then you probably made some good guesses along the way. . . . Whether this approach can be used with any confidence depends on the quality of the database used to make the predictions, of course. The alternative approach ("reverse modelling") describes the observed product and attempts to explain how it was produced, in terms of likely processes. This is the more usual approach in geology, and is the main approach described in my book.

Molinie, A.J. and Ogg, J.G., 1990. Sedimentation rate curves and discontinuities from sliding window spectral analysis of logs. *The Log Analyst*, **31** (6), 370–374.
Describes the use of a simple form of Maximum Entropy spectral analysis to detect Milankovitch cyclicity in Jurassic–Cretaceous deep marine muds, using Ocean Drilling Program GR log data. Includes many important references to related work.

Mörner, N.-A., 1994. Internal response to orbital forcing and external cyclic sedimentary sequences. In: De Boer, P.L. & Smith, D.G. (Eds.), *Orbital forcing and cyclic sequences*. Spec. Publ. IAS, **19**, 25–33.
Discusses the possible effect of Milankovitch orbital forcing on sea-level, due to tidal changes in the shape of the geoid and crustal deformation. Thus, gravitational effects may help to explain Milankovitch-related sea-level changes during periods with no significant polar ice cover. Changes in the tilt of the rotational axis (obliquity) might also change the shape of the geoid, by changing the interaction between rotational centrifugal force and the tidal forces of the Sun and Moon. It should be noted that Mörner's work has not received widespread acceptance, but nevertheless it is full of interesting and potentially revolutionary ideas.

Prokoph, A. and Agterberg, F.P., 1999. Detection of sedimentary cyclicity and stratigraphic completeness by wavelet analysis: an application to Late Albian cyclostratigraphy of the Western Canada Sedimentary Basin. *J. Sed. Research*, **69** (4), 862–875.
Uses a spectral analysis method, based on continuous wavelet transforms (also called "Morlet" analysis), of numerous gamma-ray logs to detect 100 ka-scale and longer-period rhythms in mud-dominated and sand-rich sections.

These cycles can be correlated throughout the region. Cyclic changes from wet to dry climate, probably forced by 100 ka Milankovitch orbital eccentricity cycles, are thought to have controlled the sedimentary cyclicity.

All of the deposits described in this paper formed during the early to mid Cretaceous, when there were no ice-caps – this is therefore an important advance from earlier published work on climatic cyclicity in the glaciated world.

Weedon, G.P., 1993. The recognition and stratigraphic implications of orbital-forcing of climate and sedimentary cycles. In: Wright, V.P. (Ed.), *Sedimentology review/1*. Blackwell Scientific Publications, Oxford, UK. 31–50.
An excellent introduction to the concepts of Milankovitch cyclicity, with useful references. Includes diagrams showing the orbital parameters (eccentricity, obliquity and precession) described above, together with summarised data for the last 2.5 Ma. Table 3.2 lists published examples of pre-1993 studies linking regional or global climate variations to sedimentary cylicity.

Probably the best source of up to date information on these topics is the Internet. Try using the names of the authors listed above in a search, and you ought to find something current. Of course, the Web contains a lot of unreliable information, including the often misleading statements of various religious fanatics. Good starting sites include those of the Geological Society of London, the AAPG (American Association of Petroleum Geologists) and NASA. Some useful information and comment is provided by the various environmentalist organisations (e.g. Greenpeace and the Friends of the Earth), although these obviously have a radical political agenda and therefore do not always present a balanced view.

Index

A
Aborigines 107
accommodation space 38, 39, 80, 83, 84, 85
Aegean Sea 18
Afghanistan 21
age of the Earth 6, 8, 63
air photographs 51
Alps 16
Amazon 81, 87
ammonites 74
Andes 18
apatite 63
aphelion 65
asteroid 88, 123, 124, 127
astronomical clocks 64
Atlantic Ocean 19, 69, 70
Atlantis 18

B
back-arc basins 18
Barents 108
bedding 8, 9
Bermuda Triangle 127
biostratigraphy 73
blow-out preventer 51
braided rivers 32
burrows 48

C
calcrete 29
carbon
 dating 62
 dioxide 29, 35, 88, 106
 isotopes 62, 79, 103
casing 52
Caspian Sea 127
chaos theory 125
chemical stratigraphy 89
clay 29
climate 39, 40, 81, 85, 88, 91, 105, 106, 126
coal 24, 35, 85
coastal onlap 82
colour 45
comet 88, 123, 124, 127
compaction 48
continental
 collision 13, 15, 23
 crust 10, 11, 14, 16
 drift 11, 12, 14
 shelf 20, 80
convection 10
cores 53, 112
Creation myths 2
Cretaceous 74, 88, 110, 124
cross-bedding 8, 46
cuttings 52
cycles 87, 94, 95, 96, 98, 113, 116
CycloLog 113–116
cyclostratigraphy 89

D
delta 33, 34, 35, 38
dendrochronology 59
density log 55, 56, 96, 112
deposition 10
deserts 31
dinosaurs 124
dolomite 56
drilling 51, 52, 53
dunes 31, 32, 46

E
earthquakes 14
eccentricity 65, 66, 68, 71, 98, 117, 120
evaporites 35
event stratigraphy 88

F
failed rift basins 20, 22, 110
faults 14, 15, 20, 21, 22
field-work 40, 41, 44, 48
Flandrian 107
Flood, the 2
flood-plain 1, 32, 35, 103
flute casts 35
fore-arc basins 16, 17
foreland basins 15, 16, 23, 102
Fourier Transform Analysis 90

G
gamma-ray log 54, 55, 56, 57, 91, 96, 113
Ganges River 102
geophones 50
glacier 28, 58
global warming 65, 88, 107, 127, 128
Gondwanaland 110
grain size 24, 44, 47
grass 105, 106
gravel 24, 47
gravity flows 30, 34, 35, 47
Great Rift Valley 19
greenhouse gases 65
Greenland 108
Gulf of Aden 19, 110, 111
Gulf of Mexico 80, 81, 82
Gulf of Suez 22
Gulf Stream 108, 125, 127

H
half-life 62
high-stand 85, 86
Himalayas 15, 16, 21, 23, 28, 30, 58, 101, 102, 105, 125
hominids 106
hurricanes 34, 59

I
Ice Age 6, 28, 58, 66, 69, 81, 87, 89, 107, 125
Iceland 21
image logs 57
impacts 88, 123, 124, 127
India 15, 21, 23, 30, 42, 105, 124
indigenous sediments 24, 35
Indus 58
insolation 66, 68
Iran 19, 21
island arc 16, 17

J
Japan 16, 17, 18
Jurassic 73, 74, 88, 110

L
lakes 24, 34, 35, 38, 122
lamination 46
landslides 29, 30, 59
levees 33
lime mud 44, 45
limestone 24, 29, 35, 36, 45
loess 69, 72
log trends 113, 114, 115
low-stand 83, 84, 86

M
magma 5, 13, 14, 16
magnetic
    reversal 76, 127
    stratigraphy 76, 77, 78, 103
Makran 17
Man 107, 125
Maximum Entropy analysis 91, 93, 113
maximum flooding surface 85, 86
meandering rivers 32, 33
methane hydrate 127, 128
microfossils 52, 75
mid-ocean ridge 12, 19, 22, 110
Milankovitch cycles 66, 68, 69, 72, 87, 89, 90, 98, 99, 113, 122, 126
Miocene 74, 105
Mississippi 81, 87
monsoon 105, 106
mountain building 13, 15, 23
mud 24
    log 52

N
Nepal 77, 78, 101, 102, 104
neutron log 55, 56, 96, 112
Niger delta 20, 81, 87
Nile delta 87
North Sea 20, 95
Novaya Zemlya 108

O
obduction 18, 20
obliquity 65, 71, 98, 120
Occam's Razor 123
ocean crust 11, 13, 14, 16, 17, 21, 22
oceanic anoxic events 88
Oman 18, 111
ophiolite 18
oxygen isotopes 69, 70

P
Pacific Ocean 16, 69, 79
Pakistan 9, 17, 21, 32, 41, 47, 101, 103, 106
parasequences 87, 91, 92
passive margin basins 20, 23, 24, 25
PEF log 56, 57, 112
perihelion 65, 66, 68
permeability 49
plate tectonics 10, 12, 21, 23, 39
plates 11, 12
pluvial 69
porosity 49
potassium-argon 62
precession 66, 67, 68, 98
progradation 38, 80, 91
pull-apart basins 21

R
radiometric clocks 61, 73
Red Sea 19, 22
reefs 36, 37, 39
remote sensing 50
retro-arc basins 18
rift basins 19, 22, 24, 25
ripples 33, 34, 46, 47
rivers 31, 32
rock types 5, 44, 45
roots 33, 47
rubidium-strontium 62

S
Sahara 107
salt 35
samples 49, 52
San Andreas Fault 21
sand 24, 29
    bars 31, 33, 34
Santorini 1, 18
satellite images 51
Scotland 8, 16, 31
sea level 1, 6, 40, 80, 81, 82, 83, 84, 87, 88, 107, 110, 122
seasons 64, 65
sediment supply 37, 38, 39, 40, 80, 85, 87
sedimentary
    basins 10, 14, 21, 23
    structures 46
seismic surveys 49, 50, 56, 80, 86, 87
sequence
    boundary 83, 86,
    stratigraphy 80, 83, 85, 87
shale 46
shells 36
silt 29
sliding window 91, 94, 95, 96, 97
soils 29, 103
sonic log 56
sorting 45
spectral gamma-ray log 54, 55
stable isotopes 77, 79
stalagmites 35
storms 34, 47
strata 8
stratigraphy 59
strontium isotopes 79, 110
structural dip 43
subduction 12, 13, 16, 17, 18, 22
subsidence 22, 23, 37, 38, 96
succession 37
Superposition 8
supervolcanoes 88, 125, 127

T
thermal sag basins 21, 22, 23, 110
thrusts 15, 17, 23, 102
Tibet 23, 105, 106
time scale 6, 21, 39, 74
transgressive surface 85, 86
transported sediments 24
tsunami 1, 123, 127
tufa 35
turbidites 25, 30, 35
Turkey 9, 19

U
unconformity 8, 9
Uniformitarianism 7

V
Vikings 108
volcanoes 12, 13, 14, 16, 17, 19, 30, 88

W
Walsh Power Spectra 90
water-escape 47
wireline log 54, 112, 114

Y
Yemen 19, 101, 108–122